TEST ITEMS FOR
STARR AND McMILLAN'S
HUMAN BIOLOGY
FOURTH EDITION

Larry G. Sellers
Louisiana Tech University

Allen Reich
Harvard University

Sloan D. Caldwell
Georgia College and State University

Jane B. Taylor
Northern Virginia Community College

Kendall Corbin
University of Minnesota

Tommy E. Wynn
North Carolina State University

John D. Jackson
North Hennepin Community College

p.118
p.122
p.123
p.124
p.129

p.135
p.137
p.138
p.139
p.140
p.162

p.1
p.2
p.11

p.16
p.17

p.193

p.183
p.184
p.187
p.188
p.192

p.175
p.176
p.177
p.178
p.179
p.180
p.182

BROOKS/COLE
★
™
THOMSON LEARNING

Australia • Canada • Mexico • Singapore • Spain • United Kingdom • United States

BROOKS/COLE

™

THOMSON LEARNING

Cover Design: *Denise Davidson*
Cover Photo: *PhotoDisc*

For more information about this or any other Brooks/Cole product, contact:
BROOKS/COLE
511 Forest Lodge Road
Pacific Grove, CA 93950 USA
www.brookscole.com
1-800-423-0563 (Thomson Learning Academic Resource Center)

For permission to use material from this work, contact us by
Web: www.thomsonrights.com
fax: 1-800-730-2215
phone: 1-800-730-2214

Printed in the United States of America

5   4   3   2   1

ISBN 0-534-38151-0

PREFACE

This book contains nearly 2400 test items to accompany Starr & McMillan's *Human Biology*, Fourth Edition. The test items are available on disk.

We hope you find these questions useful, and we welcome any comments that will help to improve them.

Test Items

Several instructors contributed to the test bank, including three who have extensive experience writing questions for the Educational Testing Service. Eight reviewers also have contributed valuable suggestions to help eliminate inadvertent ambiguity and to check for possible errors. The test bank represents a consensus of the kind of questions that are most suitable for students.

All questions are ranked according to level of difficulty (E = Easy, M = Moderate, and D = Difficult). Each rank is represented by about a third of the total questions. The Test Items section of this book includes the following categories:

1. Multiple-choice questions
2. Matching questions
3. Classification questions, which use the same group of answers for a series of questions.
4. Select the Exception questions, which require the student to select the exception from four or five given answers.
5. Problems, which appear only in the genetics chapters.

CONTENTS

CHAPTER 1

CONCEPTS AND METHODS IN HUMAN BIOLOGY

Multiple-Choice Questions

DNA, ENERGY, AND LIFE

E   **1.** Which is the smallest unit of life that can exist as a separate entity?
    *   a. a cell
      b. a molecule
      c. an organ
      d. a population
      e. an ecosystem

E   **2.** The capacity to do work is:
      a. matter
    *   b. energy
      c. metabolism
      d. aerobic respiration
      e. life

E   **3.** The chemical processes in the living cell are collectively called
      a. adaptation.
      b. homeostasis.
      c. evolution.
      d. respiration.
    *   e. metabolism.

E   **4.** During metabolism, ATP (adenosine triphosphate) is an energy source for the following processes:
      a. reproduction and growth
      b. reproduction and maintenance
      c. growth only
      d. growth and maintenance
    *   e. reproduction, growth, and maintenance

E   **5.** The ability to acquire, store, transfer, or utilize energy is called
      a. biochemistry.
      b. photosynthesis.
    *   c. metabolism.
      d. respiration.
      e. phosphorylation.

E   **6.** The ability to maintain a constant internal environment is
      a. metabolism.
    *   b. homeostasis.
      c. development.
      d. physiology.
      e. thermoregulation.

E   **7.** Homeostasis provides what kind of environment?
      a. positive
    *   b. constant
      c. limiting
      d. changing
      e. chemical and physical

M   **8.** The adjective that best describes homeostasis in living organisms is
      a. rigid.
      b. biological.
    *   c. dynamic.
      d. chemical.
      e. physical.

E   **9.** Each cell is able to maintain a constant internal environment. This is called
      a. metabolism.
    *   b. homeostasis.
      c. physiology.
      d. adaptation.
      e. evolution.

M   **10.** About twelve to twenty-four hours after the last meal, a person's blood sugar level normally varies from 60 to 90 milligrams per 100 milliliters of blood, though it may rise to 130 mg/100 ml after meals high in carbohydrates. That the blood sugar level is maintained within a fairly narrow range despite uneven intake of sugar is due to the body's ability to carry out
      a. adaptation.
      b. inheritance.
      c. metabolism.
    *   d. homeostasis.
      e. all of these

TWO MORE LIFE CHARACTERISTICS: REPRODUCTION AND INHERITANCE

E   **11.** Hereditary instructions must
      a. be unchanging most of the time.
      b. pass from one generation to the next.
      c. control a large number of different characteristics.
      d. provide for the rare change in instructions.
    *   e. all of these

E **12.** A mutation is a change in
   a. homeostasis.
   b. the developmental pattern in an organism.
   c. metabolism.
   * d. hereditary instructions.
   e. the life cycle of an organism.

M **13.** An adaptive trait is one that has
   a. mutated.
   * b. survival value.
   c. decreased in frequency in a population.
   d. deleterious biological effects.
   e. the potential to reduce variation.

ENERGY AND LIFE'S ORGANIZATION

M **14.** All organisms are alike in
   a. their requirements for energy.
   b. their participation in one or more nutrient cycles.
   c. their ultimate dependence on the sun.
   d. their interaction with other forms of life.
   * e. all of these

M **15.** Which of the following do not depend directly on sunlight for energy?
   I. producers
   II. consumers
   III. decomposers
   a. I only
   * b. II and III only
   c. II only
   d. III only
   e. I and III

D **16.** Which of the following would NOT be characteristic of living organisms?
   a. complex structural organization
   b. dependence on other organisms for energy and resources
   c. reproductive capacity
   * d. uniformity of size and form
   e. capacity to evolve

WHAT IS SCIENCE?

M **17.** A scientific principle is a(n)
   a. observable fact of nature.
   * b. synthesis of several explanations of many observations.
   c. scientific statement.
   d. testable hypothesis.
   e. experimental procedure.

M **18.** Of the following, which is the first explanation of a problem? It is sometimes called an "educated guess."
   a. principle
   b. law
   c. theory
   d. fact
   * e. hypothesis

E **19** Hypotheses are
   a. often in the form of a statement.
   b. often expressed negatively.
   c. sometimes crude attempts to offer a possible explanation for observations.
   d. testable predictions.
   * e. all of these

E **20.** Which statement is true about observations in the scientific process?
   a. They are made directly.
   b. They are made indirectly.
   c. Special equipment may be necessary.
   d. They may be made with an instrument such as a microscope.
   * e. all of these

E **21.** In a scientific experiment, conditions that could affect the outcome of the experiment, but do not because they are held constant, are called
   a. independent variables.
   b. dependent variables.
   * c. controlled variables.
   d. statistical variables.
   e. data set.

M **22.** To eliminate the influence of uncontrolled variables during experimentation, one should
   a. increase the sampling error as much as possible and suspend judgment.
   * b. establish a control group identical to the experimental group except for the variable being tested.
   c. use inductive reasoning to construct a hypothesis.
   e. all of these

E **23.** Which represents the lowest degree of certainty?
   * a. hypothesis
   b. conclusion
   c. fact
   d. principle
   e. theory

E   24. In order to arrive at a solution to a problem
          a scientist usually proposes and tests
          a. laws.
          b. theories.
      *   c. hypotheses.
          d. principles.
          e. facts.

E   25. The validity of scientific discoveries should
          be based on
          a. morality.
          b. aesthetics.
          c. philosophy.
          d. economics.
      *   e. none of these

E   26. Science is based on
          a. faith.
          b. authority.
      *   c. evidence.
          d. force.
          e. consensus.

A SCIENTIFIC METHOD IN ACTION: PROBING A BURNING

D   27   Which statement could be considered a
          scientific principle?
          a. The proportions of the Miss America
              contestants have been increasing over
              the last three decades.
          b. Chemistry and physics are more exact
              sciences than biology.
      *   c. A portion of sunlight consists of
              ultraviolet light.
          d. The growth of a plant is faster in a
              growth chamber than in a greenhouse.
          e. Leaves bend toward the light because
              they know light is needed to grow.

M   28. The control in an experiment
          a. makes the experiment valid.
          b. is an additional replicate for statistical
              purposes.
          c. reduces the experimental errors.
          d. minimizes experimental inaccuracy.
      *   e. allows a standard of comparison for the
              experimental group.

D   29. Which statement is false?
          a. It is easier to prove something false
              than true.
          b. Scientific experiments have limited
              applications.
          c. Experimental data are valid if they can
              be repeatedly obtained by the same
              experiment.
      *   d. Scientific conclusions are invalid if any
              step in the scientific method is omitted.
          e. Good science often uses
              experimentation.

M   30. An experimenter does all but which of the
          following?
          a. revises a hypothesis as a result of data
              collected
      *   b. manipulates dependent variables
          c. reviews other research results obtained
              by other scientists
          d. examines the effects of independent
              variables
          e. draws conclusions based only on
              appropriate experimental data

M   31. As a result of experimentation
          a. more hypotheses may be developed.
          b. more questions may be asked.
          c. a new biological principle could
              emerge.
          d. entire theories may be modified or
              discarded.
      *   e. all of these

M   32. In an experiment, the control group
          a. is not subjected to experimental error.
          b. is exposed to experimental treatments.
      *   c. is maintained under strict laboratory
              conditions.
          d. is treated exactly the same as the
              experimental group, except for the one
              independent variable.
          e. is statistically the most important part
              of the experiment.

E   33. The choice of whether a particular organism
          belongs to the experimental group or the
          control group should be based on
          a. age.
          b. size.
      *   c. chance.
          d. designation by the experimenter.
          e. sex.

M   **34.**   After an experiment is completed and the results collected, the next step is to
   a. resample the data.
   b. generalize from the conclusion.
   c. randomize the results.
*   d. organize the data.
   e. manipulate the results.

Classification Questions

Answer questions 35–38 by matching the statement to the most appropriate function, process, or trait listed below.

   a. metabolism
   b. reproduction
   c. photosynthesis
   d. growth
   e. homeostasis

M   **35.**   A process found only in plants and some bacteria

E   **36.**   Most organisms exhibit this characteristic that tends to buffer the effects of environmental change

M   **37.**   The capacity to acquire, store, and use energy

E   **38.**   Process in which one generation replaces another

Answers: 35. c   36. e   37. a
                38. b

Selecting the Exception

E   **39.**   Four of the five answers listed below are necessary characteristics to the life of an individual. Select the exception.
   a. metabolism
   b. homeostasis
   c. development
   d. heredity
*   e. diversity

E   **40.**   Four of the five answers listed below are aspects of the scientific method. Select the exception.
   a. observation
   b. hypothesis
   c. experimentation
*   d. philosophy
   e. conclusion

M   **41.**   Four of the five answers listed below are terms associated with the scientific method. Select the exception.
*   a. fact
   b. theory
   c. principle
   d. law
   e. hypothesis

M   **42.**   Four of the five answers listed below are characteristics of life. Select the exception.
*   a. ionization
   b. metabolism
   c. reproduction
   d. growth
   e. cellular organization

M   **43.**   Four of the five answers listed below are life processes that are characteristic of a normal life . Select the exception.
   a. reproduction
   b. metabolism
*   c. mutation
   d. respiration
   e. homeostasis

CHAPTER 2
THE CHEMISTRY OF LIFE

Multiple-Choice Questions

THE ATOM

M   1. Which is the smallest portion of a substance that retains the properties of an element?
    *   a. atom
       b. compound
       c. ion
       d. molecule
       e. mixture

E   2. Anything that has mass and occupies space is defined as
       a. an element.
       b. an atom.
       c. a molecule.
    *   d. matter.
       e. a compound.

E   3. The atomic number refers to the
       a. mass of an atom.
    *   b. number of protons in an atom.
       c. number of both protons and neutrons in an atom.
       d. number of neutrons in an atom.
       e. number of electrons in an atom.

M   4. Radioactive isotopes
       a. are electrically unbalanced.
       b. behave the same chemically and physically but differ biologically from other isotopes.
       c. are the same physically and biologically but differ from other isotopes chemically.
    *   d. have an excess number of neutrons.
       e. are produced when substances are exposed to radiation.

E   5. Which is NOT a compound?
       a. salt
       b. a carbohydrate
    *   c. carbon
       d. a nucleotide
       e. methane

E   6. The negative subatomic particle is the
       a. neutron.
       b. proton.
    *   c. electron.
       d. neutron and proton
       e. proton and electron

E   7. The positive subatomic particle is the
       a. neutron.
    *   b. proton.
       c. electron.
       d. neutron and proton
       e. proton and electron

E   8. The neutral subatomic particle is the
    *   a. neutron.
       b. proton.
       c. electron.
       d. neutron and proton
       e. none of these

E   9. The nucleus of an atom contains
    *   a. neutrons and protons.
       b. neutrons and electrons.
       c. protons and electrons.
       d. protons only.
       e. neutrons only.

E   10. Which components of an atom are negatively charged?
    *   a. electrons only
       b. protons only
       c. neutrons only
       d. electrons and protons
       e. protons and neutrons

E   11. Which components of an atom do not have a charge?
       a. electrons only
       b. protons only
    *   c. neutrons only
       d. electrons and protons
       e. protons and neutrons

M   12. The atomic weight (or mass) of an atom is determined by the weight of
    *   a. neutrons and protons.
       b. neutrons and electrons.
       c. protons and electrons.
       d. protons only.
       e. neutrons only.

M 13. The atomic number is determined by the number of
    a. neutrons and protons.
    b. neutrons and electrons.
    c. protons and electrons.
  * d. protons only.
    e. neutrons only.

E 14. All atoms of an element have the same number of
    a. ions.
  * b. protons.
    c. neutrons.
    d. electrons.
    e. protons and neutrons.

M 15. Radioactive isotopes have
    a. excess electrons.
    b. excess protons.
  * c. excess neutrons.
    d. insufficient neutrons.
    e. insufficient protons.

SCIENCE COMES TO LIFE: USING RADIOISOTOPES TO TRACK CHEMICALS AND SAVE LIVES

M 16. Radioactive iodine tends to concentrate in the
    a. heart.
    b. lungs.
    c. gonads.
    d. bones.
  * e. thyroid glands.

WHEN ATOM BONDS WITH ATOM

M 17. Magnesium has 12 protons. How many electrons are in its third energy level?
  * a. 2
    b. 4
    c. 6
    d. 8
    e. 10

M 18. Magnesium has 12 protons. How many electrons are in the first energy level?
  * a. 2
    b. 4
    c. 6
    d. 8
    e. 10

M 19. Magnesium has 12 protons. How many electrons are in its second energy level?
    a. 2
    b. 4
    c. 6
  * d. 8
    e. 10

M 20. Which statement is NOT true?
    a. Electrons closest to the nucleus are at the lowest energy level.
    b. No more than two electrons can occupy a single orbital.
  * c. Electrons are unable to move out of the assigned orbital space.
    d. The innermost orbital holds two electrons.
    e. At the second energy level there are four possible orbitals with a total of eight electrons.

D 21. When a molecule is excited by heat or light
    a. it may lose an electron.
    b. it may gain an electron.
  * c. an electron from an inner energy level may move to another level.
    d. an electron from an outer energy level may move to an inner level.
    e. an electron may be ejected from the nucleus of the atom.

D 22. If the atomic weight of carbon is 12 and the atomic weight of oxygen is 16, the molecular weight of glucose $C_6H_{12}O_6$ expressed in grams is
    a. 24 grams.
    b. 28 grams.
    c. 52 grams.
    d. 168 grams.
  * e. 180 grams.

E 23. Water is an example of a(n)
    a. atom.
    b. ion.
  * c. compound.
    d. mixture.
    e. element.

M 24. Which response includes the other four?
    a. atoms
  * b. molecules
    c. electrons
    d. elements
    e. protons

M   25. Which is NOT an element?
    *   a. water
        b. oxygen
        c. carbon
        d. chlorine
        e. hydrogen

M   26. A molecule is
    *   a. a combination of two or more atoms.
        b. less stable than its constituent atoms separated.
        c. electrically charged.
        d. a carrier of one or more extra neutrons.
        e. none of these

IMPORTANT BONDS IN BIOLOGICAL MOLECULES

E   27. What is formed when an atom loses or gains an electron?
        a. mole
    *   b. ion
        c. molecule
        d. bond
        e. reaction

E   28. The bond in table salt (NaCl) is
        a. polar.
    *   b. ionic.
        c. covalent.
        d. double.
        e. nonpolar.

D   29. The shape (or tertiary form) of large molecules is often controlled by what kind of bonds?
    *   a. hydrogen
        b. ionic
        c. covalent
        d. inert
        e. single

D   30. A hydrogen bond is
        a. a sharing of a pair of electrons between a hydrogen and an oxygen nucleus.
        b. a sharing of a pair of electrons between a hydrogen nucleus and either an oxygen or a nitrogen nucleus.
    *   c. an attractive force that involves a hydrogen atom and an oxygen or a nitrogen atom that are either in two different molecules or within the same molecule.
        d. none of these
        e. all of these

PROPERTIES OF WATER THAT SUPPORT LIFE

M   31. How do hydrophobic molecules interact with water?
        a. attracted to
        b. absorbed by
    *   c. repelled by
        d. mixed with
        e. polarized by

D   32. Water is an excellent solvent because
        a. it forms spheres of hydration around charged substances and can form hydrogen bonds with many nonpolar substances.
        b. it has a high heat of fusion.
        c. of its cohesive properties.
        d. it is a liquid at room temperature.
    *   e. all of these

D   33. In a lipid bilayer, _____ tails point inward and form a region that excludes water.
        a. acidic
        b. basic
        c. hydrophilic
    *   d. hydrophobic
        e. none of these

D   34. Glucose dissolves in water because it
        a. ionizes.
        b. is a polysaccharide.
    *   c. is polar and forms many hydrogen bonds with the water molecules.
        d. has a very reactive primary structure.
        e. none of these

ACIDS, BASES, AND BUFFERS

M   35. Which of the following is a free (unbound) proton?
    *   a. hydrogen ion
        b. acid
        c. base
        d. hydroxyl ion
        e. acceptor

D   36. Sodium chloride (NaCl) in water could be described by any of the following EXCEPT:
        a. Na^+ and Cl^- form
        b. a solute
        c. ionized
    *   d. forms spheres of hydration
        e. dissolved

M   37. A salt will dissolve in water to form
   - a. acids.
   - b. gases.
   - * c. ions.
   - d. bases.
   - e. polar solvents.

M   38. A reaction of an acid and a base will produce water and
   - a. a buffer.
   - * b. a salt.
   - c. gas.
   - d. solid precipitate.
   - e. solute.

M   39. Which of the following would NOT be used in connection with the word *acid*?
   - a. excess hydrogen ions
   - b. contents of the stomach
   - * c. magnesium hydroxide
   - d. HCl
   - e. pH less than 7

M   40. Cellular pH is kept near a value of 7 because of
   - a. salts.
   - * b. buffers.
   - c. acids.
   - d. bases.
   - e. water.

M   41. A pH of 10 is how many times as basic as a pH of 7?
   - a. 2
   - b. 3
   - c. 10
   - d. 100
   - * e. 1,000

M   42. A solution with a pH of 8 has how many times fewer hydrogen ions than a solution with a pH of 6?
   - a. 2
   - b. 4
   - c. 10
   - * d. 100
   - e. 1,000

ORGANIC COMPOUNDS: BUILDING ON CARBON ATOMS

E   43. The three most common atoms in your body are
   - * a. hydrogen, oxygen, and carbon.
   - b. carbon, hydrogen, and nitrogen.
   - c. carbon, nitrogen, and oxygen.
   - d. nitrogen, hydrogen, and oxygen.
   - e. carbon, oxygen, and sulfur.

E   44. Carbon usually forms how many bonds with other atoms?
   - a. 2
   - b. 3
   - * c. 4
   - d. 5
   - e. 6

E   45. The atom diagnostically associated with organic compounds is
   - * a. carbon.
   - b. oxygen
   - c. nitrogen.
   - d. sulfur.
   - e. hydrogen.

M   46. Which are NOT macromolecules?
   - a. proteins
   - b. starches
   - * c. nucleotides
   - d. lipids
   - e. nucleic acids

D   47. Which compound is hydrophobic?
   - a. ethyl alcohol
   - b. simple sugar
   - * c. hydrocarbon
   - d. glycerol
   - e. amino acid

M   48. An —OH group is a(n) _____ group.
   - a. carboxyl
   - * b. hydroxyl
   - c. amino
   - d. methyl
   - e. ketone

M   49. A —CH_3 group is a(n) _____ group.
   - a. carboxyl
   - b. hydroxyl
   - c. amino
   - * d. methyl
   - e. ketone

M   **50.** An —NH$_2$ group is a(n) _____ group.
- a. carboxyl
- b. hydroxyl
- * c. amino
- d. methyl
- e. ketone

M   **51.** A —COOH group is a(n) _____ group.
- * a. carboxyl
- b. hydroxyl
- c. amino
- d. methyl
- e. ketone

M   **52.** The formation of large molecules from small repeating units is known as what kind of reaction?
- a. oxidation
- b. reduction
- * c. condensation
- d. hydrolysis
- e. decarboxylation

M   **53.** The breakdown of large molecules by the enzymatic addition of water is an example of what kind of reaction?
- a. oxidation
- b. reduction
- c. condensation
- * d. hydrolysis
- e. decarboxylation

M   **54.** Which reaction results in the breakdown of a chemical into simpler substances?
- a. synthesis
- * b. hydrolysis
- c. condensation
- d. polymerization

E   **55.** Which substance is the most common in cells?
- a. carbohydrates
- b. salts and minerals
- c. proteins
- d. fats
- * e. water

CARBOHYDRATES

M   **56.** Which of the following includes all the others?
- a. sucrose
- b. glucose
- c. cellulose
- d. glycogen
- * e. carbohydrate

E   **57.** Which is a "building block" of carbohydrates?
- a. glycerol
- b. nucleotide
- c. simple sugar
- d. monosaccharide
- * e. simple sugar or monosaccharide

M   **58.** Which of the following is composed of a 1:2:1 ratio of carbon to hydrogen to oxygen?
- * a. carbohydrate
- b. protein
- c. lipid
- d. nucleic acid
- e. steroid

M   **59.** Which is NOT a monosaccharide?
- a. glucose
- b. fructose
- c. deoxyribose
- * d. starch
- e. ribose

M   **60.** Cellulose is
- * a. a material found in cell walls.
- b. a component of cell membranes.
- c. a plant protein.
- d. formed by photosynthesis.
- e. the most complex of the organic compounds.

D   **61.** Monosaccharides are characterized by all EXCEPT which of the following?
- a. a carboxyl group
- b. carbon, hydrogen, and oxygen in a 1:2:1 ratio
- c. a molecule of three to seven carbon atoms
- d. possession of one or more hydroxyl groups
- * e. the presence of glycerol and fatty acids

M **62.** Fructose and glucose are
  a. isotopes.
  b. monosaccharides.
  c. disaccharides.
  d. six-carbon sugars.
  * e. monosaccharides and six-carbon sugars

M **63.** Fructose and glucose are
  a. hexoses.
  b. structurally different.
  c. monosaccharides.
  d. simple sugars.
  * e. all of these

M **64.** Glucose and ribose
  a. have the same number of carbon atoms.
  b. have the same structural formulas.
  c. are the two components of sucrose.
  * d. are monosaccharides.
  e. are molecules whose atoms are arranged the same way.

M **65.** Sucrose is composed of
  a. two molecules of fructose.
  b. two molecules of glucose.
  * c. a molecule of fructose and a molecule of glucose.
  d. a molecule of fructose and a molecule of galactose.
  e. two molecules of fructose.

M **66.** The combination of glucose and galactose forms
  a. fructose.
  b. maltose.
  * c. lactose.
  d. sucrose.
  e. mannose.

E **67.** Plants store their excess carbohydrates in the form of
  * a. starch.
  b. glycogen.
  c. glucose.
  d. cellulose.
  e. fats.

M **68.** Glycogen is a polysaccharide used for energy storage by
  * a. animals.
  b. plants.
  c. algae.
  d. bacteria.
  e. animals and algae

LIPIDS

M **69.** Triglycerides are
  a. carbohydrates.
  b. nucleotides.
  c. proteins.
  * d. fats.
  e. amino acids.

M **70.** Oils are
  a. liquid at room temperatures.
  b. unsaturated fats.
  c. found only in animals.
  d. complex carbohydrates.
  * e. liquid at room temperatures and unsaturated fats.

E **71.** Which of the following are lipids?
  a. steroids
  b. triglycerides
  c. oils
  d. waxes
  * e. all of these

M **72.** An example of a saturated fat is
  a. olive oil.
  b. corn oil.
  * c. butter.
  d. oleo.
  e. soybean oil.

M **73.** Lipids
  * a. serve as food reserves in many organisms.
  b. include cartilage and chitin.
  c. include fats that are broken down into one fatty acid molecule and three glycerol molecules.
  d. are composed of monosaccharides.
  e. none of these

M **74.** Plasma membranes are characterized by the presence of
  a. triglycerides.
  * b. phospholipids.
  c. unsaturated fats.
  d. steroid hormones.
  e. fatty acids.

M **75.** All steroids have
  a. the same number of double bonds.
  b. the same position of double bonds.
  * c. four rings of carbon to which are attached other atoms.
  d. the same functional groups.
  e. the same number and same positions of double bonds

M   **76.** Steroids are
    a. compounds that are related to lipids.
    b. sex hormones.
    c. components of membranes.
    d. troublesome on walls of arteries.
*   e. all of these

AMINO ACIDS AND THE PRIMARY STRUCTURE OF PROTEINS

D   **77.** Which element is NOT characteristic of the primary structure of proteins?
    a. sulfur ✓
    b. carbon ✓
*   c. phosphorus
    d. oxygen ✓
    e. nitrogen

$$R - \overset{\overset{H}{|}}{\underset{\underset{NH_2}{|}}{C}} - COOH$$

M   **78.** Proteins may function as
    a. structural units.
    b. hormones.
    c. storage molecules.
    d. transport molecules.
*   e. all of these

E   **79.** Amino acids are the building blocks for
*   a. proteins.
    b. steroids.
    c. lipids.
    d. nucleic acids.
    e. carbohydrates.

E   **80.** What kind of bond exists between two amino acids in a protein?
*   a. peptide
    b. ionic
    c. hydrogen
    d. amino
    e. sulfhydroxyl

E   **81.** The sequence of amino acids is the _____ structure of proteins.
*   a. primary
    b. secondary
    c. tertiary
    d. quaternary
    e. stereo

E   **82.** Amino acids are linked by what kind of bonds to form the primary structure of a protein?
    a. disulfide
    b. hydrogen
    c. ionic
*   d. peptide
    e. none of these

HOW DOES A PROTEIN GET ITS THREE-DIMENSIONAL STRUCTURE

M   **83.** The secondary structure of proteins is
    a. helical.
    b. sheetlike.
    c. globular.
    d. the sequence of amino acids.
*   e. helical or sheetlike.

E   **84.** Which of the following exhibits fourth level structure?
    a. amino acids
    b. lipids
*   c. hemoglobin
    d. glycogen
    e. polysaccharide

M   **85.** A glycoprotein is a combination of protein and
    a. heme.
*   b. oligosaccharides.
    c. collagen.
    d. fatty acids.
    e. nucleic acids.

M   **86.** The disruption of a protein's three-dimensional shape is called
    a. condensation.
    b. hydrolysis.
    c. ionization.
*   d. denaturation.
    e. oxidation.

NUCLEOTIDES AND NUCLEIC ACIDS

M   **87.** Which of the following is NOT found in every nucleic acid?
*   a. ribose
    b. phosphate group
    c. purine
    d. pyrimidine
    e. all of these are characteristic of every nucleotide

M   **88.** Nucleotides are the building blocks for
    a. proteins.
    b. steroids.
    c. lipids.
*   d. ATP, NAD^+, and FAD.
    e. carbohydrates.

M   89. The nucleotide most closely associated with energy is
         a. cyclic AMP.
         b. FAD.
         c. NAD.
    *   d. ATP.
         e. all of these

M   90. Nucleotides contain what kind of sugars?
         a. three-carbon
         b. four-carbon
    *   c. five-carbon
         d. six-carbon
         e. seven-carbon

M   91. DNA
         a. is one of the adenosine phosphates.
         b. is one of the nucleotide coenzymes.
    *   c. contains protein-building instructions.
         d. all of these
         e. none of these

Matching Questions

M   92. Choose the one most appropriate answer for each.
         1 ____ enzyme
         2 ____ glucose
         3 ____ nucleotide coenzymes
         4 ____ phospholipids

         A. a six-carbon sugar
         B. energy carrier such as NAD and FAD
         C. principal components of cell membranes
         D. speed up metabolic reactions
         E. DNA and RNA

Answers: 1. D   2. A   3. B
              4. C

Classification Questions

The various energy levels in an atom of magnesium have different numbers of electrons. Use the following numbers to answer questions 93-95.

         a. 1
         b. 2
         c. 3
         d. 6
         e. 8

D   93. Number of electrons in the first energy level

D   94. Number of electrons in the second energy level

D   95. Number of electrons in the third energy level

Answers: 93. b   94. e   95. b

The following are types of chemical bonds. Answer questions 96-100 by matching the statement with the most appropriate bond type.

         a. hydrogen
         b. ionic
         c. covalent
         d. disulfide
         e. peptide

M   96. The bond between the atoms of table salt

M   97. The bond type holding several molecules of water together

M   98. The bond between the oxygen atoms of gaseous oxygen

M   99. The bond that breaks when salts dissolve in water

M   100. Atoms connected by this kind of bond share electrons

Answers: 96. b   97. a   98. c
              99. b   100. c

The following are chemical functional groups that may be part of a biologically active molecule. Answer questions 101-111 by matching the statement with the most appropriate group.

a. —COOH
b. —CH₃
c. —NH₂
d. —OH
e.   >C＝O
f.
$$\begin{array}{c} O \\ \| \\ —P—O \\ | \\ O \end{array}$$

g. —CHO

E **101.** The amino group

E **102.** The carboxyl group

M **103.** The group that is acidic

M **104.** The group that occurs repeatedly in sugars; composed of two elements

E **105.** The methyl group

E **106.** The hydroxyl group

E **107.** The ketone group

M **108.** The group on the amino-terminal end of proteins

M **109.** The group on the carboxyl-terminal end of proteins

D **110.** A group composed of three different elements; found in sugars

M **111.** The group typical of energy carriers such as ATP

Answers: 101. c   102. a   103. a
           104. d   105. b   106. d
           107. e   108. c   109. a
           110. g   111. f

The following are basic building blocks of biopolymers. Answer questions 112-119 by matching the statement with the most appropriate building block.

a. amino acids
b. glucose
c. glycerol
d. fatty acids
e. nucleotides

E **112.** The basic unit of proteins

E **113.** The basic unit of DNA

E **114.** The basic unit of messenger RNA

E **115.** The basic unit of cellulose

E **116.** The basic unit of glycogen

E **117.** The basic unit of starch

M **118.** The "building block" unit of a polypeptide chain

M **119.** Which two units combine in various ways to form lipids?
a. a and b
b. a and c
c. b and c
d. b and d
e. c and d

Answers: 112. a   113. e   114. e
           115. b   116. b.   117. b.
           118. a   119. e

Selecting the Exception

D **120.** Four of the five answers listed below possess electrons in the third energy level. Select the exception.
a. sodium
b. magnesium
c. chlorine
* d. nitrogen
e. sulfur

D **121.** Four of the five answers listed below are related by a unifying characteristic. Select the exception.
a. ionic bond
b. covalent bond
c. polar bond
d. hydrogen bond
* e. cluster of nonpolar groups

D **122.** Four of the five answers listed below are alkaline (pH above 7). Select the exception.
   a. milk of magnesia
   b. household ammonia
   c. Tums
   d. phosphate detergent
   * e. cola soft drink

D **123.** Four of the five answers listed below are acidic (pH below 7). Select the exception.
   a. vinegar
   b. soft drink
   * c. soap
   d. lemon juice
   e. beer

M **124.** Four of the five answers listed below are positively charged ions. Select the exception.
   a. potassium ion
   b. hydrogen ion
   c. calcium ion
   d. magnesium ion
   * e. chlorine ion

M **125.** Four of the five answers listed below are characteristics of water. Select the exception.
   a. stabilize temperature
   b. common solvent
   c. cohesion and surface tension
   * d. produce salts
   e. change shape of hydrophilic and hydrophobic substances

D **126.** Four of the five answers listed below are related by a common chemical similarity. Select the exception.
   a. cellulose
   * b. hydrochloric acid
   c. amino acid
   d. protein
   e. nucleic acid

M **127.** Four of the five answers listed below are related as members of the same group. Select the exception.
   a. glucose
   b. fructose
   * c. cellulose
   d. ribose
   e. deoxyribose

D **128.** Four of the five answers listed below are related as members of the same group. Select the exception.
   a. lactose
   b. sucrose
   c. maltose
   d. table sugar
   * e. fructose

D **129.** Four of the five answers listed below are carbohydrates. Select the exception.
   * a. glycerol
   b. cellulose
   c. starch
   d. sucrose
   e. glycogen

D **130.** Four of the five answers listed below are lipids. Select the exception.
   a. triglyceride
   b. wax
   c. cutin
   * d. insulin
   e. steroid

M **131.** Three of the four answers listed below are saturated fats. Select the exception.
   a. butter
   b. bacon
   * c. peanut oil
   d. animal fat

D **132.** Four of the five answers listed below are amino acids. Select the exception.
   a. glycine
   * b. adenine
   c. phenylalanine
   d. valine
   e. tyrosine

D **133.** Four of the five answers listed below are functional groups. Select the exception.
   * a. R group
   b. amino group
   c. carboxyl group
   d. hydroxyl group
   e. aldehyde group

CHAPTER 3

CELLS

Multiple-Choice Questions

CELLS FILL"D WITH JUICES

E **1.** The first cell that was seen under a microscope was a
*   a. cork cell.
     b. blood cell.
     c. sperm cell.
     d. skin cell.
     e. root tip cell.

E **2.** Which of the following is most accurate regarding cells?
     a. All cells have a nucleus.
     b. All cells divide by meiosis.
*   c. All living organisms are made up of cells.
     d. Cells arise through spontaneous generation.
     e. Growth is solely the result of cell division.

E **3.** Who is generally given credit for seeing cells for the first time and naming them?
*   a. Robert Hooke.
     b. Robert Brown.
     c. Theodor Schwann and Matthias Schleiden.
     d. Rudolf Virchow.
     e. Antony van Leeuwenhoek.

M **4.** The resolution of small details by a light microscope is limited by the
     a. vision of the human viewer.
     b. power of the lenses
     c. size of the specimen
*   d. properties of the light waves
     e. stains used in preparation of the specimen

E **5.** The maximum power of magnification of a light microscope is about __?__ times.
     a. 500.
     b. 1,000.
*   c. 2,000.
     d. 4,000.
     e. 10,000.

E **6.** The highest magnification generally used to study cells is provided by the
*   a. transmission electron microscope.
     b. compound light microscope.
     c. phase contrast microscope.
     d. scanning electron microscope.
     e. binocular dissecting microscope.

AN INTRODUCTION TO CELLS

E **7.** Which of the following statements about cells is NOT true?
     a. Most cells cannot be seen without the aid of a microscope.
     b. There are more kinds of eukaryotic cells than prokaryotic cells.
     c. Most cell have an organized nucleus.
     d. Some cells are large enough to see without magnification.
*   e. The plasma membrane isolates the cell's interior from the exterior.

M **8.** Which of the following is an example of a prokaryotic cell?
     a. algae
*   b. bacteria
     c. human
     d. plant
     e. fungus

EUKARYOTIC CELLS AND THEIR COMPONENTS

M **9.** These are the primary cellular sites for the production of proteins.
     a. Golgi bodies
*   b. ribosomes
     c. mitochondria
     d. lysosomes
     e. smooth endoplasmic reticula

M **10.** These are the primary structures for the packaging of cellular secretions for export from the cell.
*   a. Golgi bodies
     b. ribosomes
     c. mitochondria
     d. lysosomes
     e. endoplasmic reticula

M 11. These contain enzymes used in the breakdown of glucose and generation of ATP.
     a. Golgi bodies
     b. ribosomes
  * c. mitochondria
     d. lysosomes
     e. endoplasmic reticula

E 12. Which of these is the localized site for DNA in the cell?
     a. ribosomes
     b. Golgi body
  * c. nucleus
     d. mitochondria
     e. vesicles

E 13. Which of these is the site for protein modification and lipid synthesis?
     a. nucleus
     b. cytoskeleton
     c. mitochondria
  * d. endoplasmic reticulum
     e. all of these

E 14. These are responsible for cell shape, internal organization, and movements.
     a. ribosomes
  * b. cytoskeleton
     c. vesicles
     d. Golgi bodies
     e. endoplasmic reticulum

THE CYTOSKELETON

M 15. Structural features that influence the shapes of cells are
     a. plastids.
     b. vacuoles.
     c. microvilli.
     d. nucleoli.
  * e. microfilaments.

M 16. Organelles used to move chromosomes are the
     a. cilia.
     b. flagella.
  * c. microtubules.
     d. microfilaments.
     e. Golgi apparatuses.

E 17. The organelle that is compared to a whip is
     a. microfilament.
     b. basal body.
     c. microvillus.
  * d. flagellum.
     e. microtubule.

E 18. A 9+2 array refers to
     a. microtubules.
     b. Golgi bodies.
     c. ribosomes.
     d. cilia.
  * e. both microtubules and cilia.

THE PLASMA MEMBRANE: A "FLUID MOSAIC"

M 19. The membrane proteins responsible for distinguishing "self" from "nonself" are which of the following?
     a. channel
  * b. recognition
     c. receptor
     d. carrier
     e. mosaic

E 20. The special functions performed by membranes are the result of
     a. the number of layers of phospholipids present.
     b. concentration gradients of the interior contents.
     c. lipids embedded in the protein layer.
     d. steroids.
  * e. proteins in the lipid bilayer.

M 21. The phospholipid molecules of most membranes have
     a. a hydrophobic head and a hydrophilic tail.
     b. a hydrophobic head and a hydrophobic tail.
     c. a hydrophobic head and two hydrophobic tails.
  * d. a hydrophilic head and two hydrophobic tails.
     e. none of these

E 22. Hydrophobic reactions of phospholipids may produce clusters of their fatty acid tails, which form
  * a. a lipid bilayer.
     b. hydrolysis of the fatty acids.
     c. a protein membrane.
     d. a cytoskeleton.
     e. a nonpolar membrane.

M **23.** Unsaturated tails of lipids
   a. are hydrophilic.
   b. are unstable and tend to break apart.
 * c. have kinks in them and lessen the interaction between adjacent fat
   d. will break whenever exposed to phosphate ions.
   e. all of these

M **24.** The relative impermeability of membranes to water-soluble molecules is a result of the
   a. nonpolar nature of water molecules.
   b. presence of large proteins that extend through both sides of membranes.
   c. presence of inorganic salt crystals scattered through some membranes.
   d. presence of cellulose and chemicals such as cutin, lignin, pectin, and suberin in the membranes.
 * e. presence of phospholipids in the lipid bilayer.

THE CYTOMEMBRANE SYSTEM

E **25.** Organelle composed of a system of canals, tubes, and sacs that transport molecules inside the cytoplasm are
   a. Golgi body.
   b. ribosome.
   c. mitochondrion.
   d. lysosome.
 * e. endoplasmic reticulum.

M **26.** Which of the following consists of two subunits plus RNA and protein?
   a. Golgi
   b. mitochondria
   c. chloroplasts
 * d. ribosomes
   e. endoplasmic reticula

E **27.** These are sometimes referred to as rough or smooth depending on the structure.
   a. Golgi bodies
   b. ribosomes
   c. mitochondria
   d. lysosomes
 * e. endoplasmic reticula

M **28.** These contain enzymes and are the main organelles of intracellular digestion.
   a. Golgi bodies
   b. ribosomes
   c. mitochondria
 * d. lysosomes
   e. endoplasmic reticula

D **29.** Animal cells dismantle and dispose of intracellular waste materials by
   a. using centrally located vacuoles.
 * b. several lysosomes fusing with a sac that encloses the wastes.
   c. microvilli packaging and exporting the wastes.
   d. mitochondrial breakdown of the wastes.
   e. all of these

M **30.** The organelle that degrades potentially harmful hydrogen peroxide to harmless substances is the
   a. lysosome.
 * b. peroxisome.
   c. mitochondria.
   d. rough ER.
   e. Golgi.

D **31.** Peroxisomes would most likely be involved in the metabolism of
   a. glucose.
 * b. alcohol.
   c. white blood cells.
   d. pesticides.
   e. nucleoli.

MOVING SUBSTANCES BY DIFFUSION AND OSMOSIS

D **32.** Which statement is NOT true?
   a. Membranes are often perforated by proteins that extend through both sides of the membrane.
   b. Some membranes have proteins with channels or pores that allow for the passage of hydrophilic substances.
 * c. Hydrophilic substances have an easier time passing through membranes than hydrophobic substances do.
   d. The current concept of a membrane can be best summarized by the fluid mosaic model.
   e. The lipid bilayer serves as a hydrophobic barrier between two fluid regions.

D **33.** The movement of water through a membrane is dependent on
 * a. the concentration of solute.
   b. channel proteins.
   c. the extent of packing of the phospholipids.
   d. active transport
   e. endocytosis.

D   34. The rate of diffusion through a semipermeable membrane will be lowest when the
   a. differences in concentration on either side are the greatest.
   b. temperature is raised to near boiling.
   * c. differences in concentration on either side are the least.
   d. actions of active transport override diffusion.
   e. inside is hypertonic to the outside.

M   35. In simple diffusion
   a. the rate of movement of molecules is controlled by temperature and pressure.
   b. the movement of individual molecules is random.
   c. the movement of molecules of one substance is independent of the movement of any other substance.
   d. the net movement is away from the region of highest concentration.
   * e. all of these

M   36. A single-celled freshwater organism, such as a protistan, is transferred to salt water. Which of the following is likely to happen?
   a. The cell bursts.
   b. Salt is pumped out of the cell.
   * c. The cell shrinks.
   d. Enzymes flow out of the cell.
   e. all of these

M   37. Which statement is true?
   a. A cell placed in an isotonic solution will swell.
   * b. A cell placed in a hypotonic solution will swell.
   c. A cell placed in a hypotonic solution will shrink.
   d. A cell placed in a hypertonic solution will remain the same size.
   e. A cell placed in a hypotonic solution will remain the same size.

M   38. A red blood cell will swell and burst when placed in which of the following kinds of solution?
   * a. hypotonic
   b. hypertonic
   c. isotonic
   d. any of the above
   e. none of these

M   39. If a plant cell is placed in a hypotonic solution
   a. the entire cell will not swell or shrink.
   b. the entire cell will shrink.
   c. the turgor pressure will increase.
   d. the cell wall prevents the cell from exploding.
   * e. the turgor pressure will increase but the cell wall will prevent the cell from exploding.

OTHER WAYS SUBSTANCES CROSS MEMBRANES

M   40. Which of the following is NOT a form of active transport?
   a. sodium-potassium pump
   b. endocytosis
   c. exocytosis
   * d. diffusion
   e. none of these

M   41. Phagocytosis is a type of
   a. exocytosis.
   b. osmosis.
   c. cell membrane model.
   * d. endocytosis.
   e. diffusion.

E   42. Movement of a molecule against a concentration gradient is
   a. simple diffusion.
   b. facilitated diffusion.
   c. osmosis.
   * d. active transport.
   e. passive transport.

M   43. The method of movement that requires the expenditure of ATP molecules is
   a. simple diffusion.
   b. facilitated diffusion.
   c. osmosis.
   * d. active transport.
   e. passive transport.

M   44. White blood cells use _____ to get rid of foreign particles in the blood.
   a. simple diffusion
   b. bulk flow
   c. osmosis
   * d. phagocytosis
   e. facilitated diffusion

E   45. The accumulation of a particular material inside a cell in greater amounts than exist outside the cell is possible due to
    a. simple diffusion.
    b. facilitated diffusion.
    c. osmosis.
    * d. active transport.
    e. passive transport.

M   46. The carrier molecules used in active transport are
    a. calcium ions in the calcium pump.
    * b. membrane proteins.
    c. ATP molecules.
    d. carbohydrates.
    e. lipids.

E   47. Cells active in secreting enzymes would likely exhibit a higher than usual amount of
    a. osmosis.
    * b. exocytosis.
    c. lipid bilayers in the plasma membranes.
    d. endocytosis.
    e. receptor proteins.

THE NUCLEUS

E   48. An organelle found in the nucleus is a
    a. plastid.
    b. vacuole.
    c. microvillus.
    * d. nucleolus.
    e. basal body.

M   49. Which is NOT found as a part of all cells?
    a. cell membrane
    * b. nucleus
    c. ribosomes
    d. DNA
    e. RNA

M   50. Which of the following is NOT true of the nuclear envelope?
    * a. single lipid bilayer
    b. continuous with endoplasmic reticulum
    c. possesses pores
    d. controls passage into and out of nucleus
    e. separates DNA from cytoplasm

THE MITOCHONDRION

M   51. These are the primary cellular sites for the transfer of energy from carbohydrates.
    a. Golgi bodies
    b. ribosomes
    * c. mitochondria
    d. lysosomes
    e. endoplasmic reticula

D   52. Energy stored in which of the following molecules is converted by mitochondria to a form usable by the cell?
    a. water
    * b. carbon compounds
    c. NAD
    d. ATP
    e. carbon dioxide

M   53. The interior surface area of mitochondria is greatly increased by
    a. plastids.
    * b. cristae.
    c. centrioles
    d. nucleoli.
    e. microfilaments.

METABOLISM: DOING CELLULAR WORK

M   54. Which reaction is NOT an energy-releasing reaction?
    * a. protein synthesis
    b. digestion
    c. fire
    d. respiration
    e. movement

M   55. During enzyme catalyzed reactions, *substrate* is a synonym for
    a. end products.
    b. by-products.
    c. enzymes.
    * d. reactants.
    e. all of these

D   56. Which of the following is capable of enzymatic activity?
    a. lipids
    * b. proteins
    c. carbohydrates
    d. minerals
    e. vitamins

M 57. Which of the following is NOT true of enzyme behavior?
   a. Enzyme shape may change during catalysis.
   b. Enzymes cannot make something happen that would not happen on its own.
   c. All enzymes have an active site where substrates are temporarily bound.
 * d. Each enzyme can catalyze a wide variety of different reactions.
   e. Enzymes speed up reactions.

E 58. Enzymes
   a. are very specific.
   b. act as catalysts.
   c. are organic molecules.
   d. have special shapes that control their activities.
 * e. all of these

M 59. Enzymes
   a. control the speed of a reaction.
   b. change shapes to facilitate certain reactions.
   c. act on substrates.
   d. may require cofactors.
 * e. all of these

D 60. Which of the following substances would be unlikely to function as a coenzyme?
   a. a water-soluble vitamin
   b. an iron ion
 * c. glucose
   d. NAD^+
   e. a magnesium ion

M 61. Enzymatic reactions can be controlled by
   a. the amount of substrates available.
   b. the concentration of products.
   c. temperature.
   d. modification of reactive sites by substances that fit into the enzyme and, later, their reactive site.
 * e. all of these

M 62. A molecule that gives up an electron becomes
   a. ionized only
   b. oxidized only
   c. reduced only
 * d. ionized and oxidized
   e. oxidized and reduced

M 63. The removal of electrons from a compound is known as
   a. dehydration.
 * b. oxidation.
   c. reduction.
   d. phosphorylation.
   e. a nonreversible chemical reaction.

M 64. When NAD^+ combines with hydrogen, the NAD^+ becomes
 * a. reduced.
   b. oxidized.
   c. phosphorylated.
   d. denatured.
   e. none of these

MAKING ATP: THE FIRST TWO STAGES

M 65. ATP acts as what type of agent in almost all metabolic pathways?
 * a. energy transfer
   b. feedback
   c. catalytic
   d. synthetic
   e. enzymatic

E 66. ATP contains
   a. alanine.
   b. arginine.
 * c. ribose.
   d. tyrosine.
   e. glucose.

E 67. ATP contains
 * a. adenine.
   b. cytosine.
   c. uracil.
   d. thymine.
   e. guanine.

M 68. When molecules are broken apart in respiration
   a. the heat produced is used to drive biological reactions.
   b. the oxygen in the compounds that are broken apart is used as an energy source.
 * c. the energy released in respiration is channeled into molecules of ATP.
   d. ATP is converted into ADP.
   e. ADP is released as a waste product.

M   69. Which of the following has the greatest
        amount of energy?
        a. cAMP
        b. ADP
        c. ATP
    *   d. glucose
        e. NADPH

E   70. Humans derive most of their energy from
        a. fats.
    *   b. carbohydrates.
        c. proteins.
        d. nucleotides.
        e. steroids.

E   71. ATP is
    *   a. the energy currency of a cell.
        b. produced by the destruction of ADP.
        c. expended in the process of respiration
        d. produced during the phosphorylation of
           any organic compound.
        e. a vitamin.

M   72. ATP
        a. can be produced by endocytosis.
        b. is the by-product of the biosynthesis of
           organic compounds such as glucose.
        c. is produced in greatest amounts during
           anaerobic respiration.
    *   d. is released in aerobic respiration.
        e. all of these

E   73. Which liberates the most energy in the form
        of ATP?
    *   a. aerobic respiration
        b. anaerobic respiration
        c. alcoholic fermentation
        d. lactate fermentation
        e. All liberate the same amount, but
           through different means.

D   74. Glycolysis depends upon a continuous
        supply of
        a. NADP.
        b. pyruvate.
    *   c. glucose.
        d. NADH.
        e. H_2O.

E   75. Glycolysis
        a. occurs in the mitochondria.
        b. happens to glucose only.
        c. results in the production of pyruvate.
        d. occurs in the cytoplasm.
    *   e. occurs in the cytoplasm and results in
           the production of pyruvate.

E   76. The ultimate source of energy for living
        things is the
        a. Krebs cycle.
        b. fossil fuels.
    *   c. sun.
        d. glycolysis.
        e. aerobic respiration.

D   77. Before a glucose molecule can be broken
        down to release energy
        a. one ATP molecule must be added to
           glucose.
    *   b. two phosphate groups must be attached
           to glucose.
        c. three ATP molecules must be added to
           glucose.
        d. one ATP molecule must be taken away
           from glucose.
        e. two ATP molecules must be taken
           away from glucose.

D   78. The amount of energy released from a
        glucose molecule is dependent on what
        happens to
        a. carbon atoms.
        b. oxygen atoms.
    *   c. hydrogen atoms.
        d. phosphorus atoms.
        e. water molecules.

M   79. The end result of glycolysis is
        a. acetyl CoA.
        b. oxaloacetate.
    *   c. pyruvate.
        d. citrate.
        e. acetyl CoA and oxaloacetate.

M   80. How many ATP molecules (net yield) are
        produced per molecule of glucose degraded
        during glycolysis?
        a. 1
    *   b. 2
        c. 4
        d. 36
        e. 38

D   81. The conversion of PGAL to pyruvate
        involves
        a. anaerobic respiration.
        b. photophosphorylation.
        c. the electron transport chain.
    *   d. substrate-level phosphorylation.
        e. the Krebs cycle.

M 82. In the breakdown of glucose, the compound
        formed after two phosphorylation reactions
        is split into two three-carbon compounds.
        The three-carbon compound is named
    *   a. phosphoglyceraldehyde (PGAL).
         b. pyruvate.
         c. acetyl CoA.
         d. lactate.
         e. acetaldehyde.

D 83. Which is capable of being reduced during
        both glycolysis and the Krebs cycle?
    *   a. NAD$^+$
         b. FAD$^+$
         c. ADP
         d. NADH
         e. NADP$^+$

E 84. The Krebs cycle takes place in the
         a. ribosomes.
         b. cytoplasm.
         c. nucleus.
    *   d. mitochondria.
         e. chloroplasts.

E 85. Pyruvate can be regarded as the end product
        of
    *   a. glycolysis.
         b. acetyl CoA formation.
         c. fermentation.
         d. the Krebs cycle.
         e. electron transport.

M 86. The breakdown of pyruvate in the Krebs
        cycle results in the release of
         a. energy.
         b. carbon dioxide.
         c. oxygen.
         d. hydrogen.
    *   e. all of these except oxygen.

M 87. What is the intermediate compound
        produced in the Krebs cycle to which the
        acetyl group becomes attached?
         a. pyruvate
         b. acetyl CoA
         c. fructose bisphosphate
    *   d. oxaloacetate
         e. citrate

D 88. During which phase of aerobic respiration is
        ATP produced directly by substrate-level
        phosphorylation?
         a. glucose formation
         b. lactate production
         c. acetyl CoA formation
    *   d. the Krebs cycle

M 89. Which of these serves in the transition from
        glycolysis to the Krebs cycle?
    *   a. acetyl CoA formation
         b. conversion of PGAL to PGA
         c. regeneration of reduced NAD$^+$
         d. oxidative phosphorylation
         e. substrate-level phosphorylation

THE THIRD STAGE OF THE AEROBIC PATHWAY

M 90. Aerobes use _____ as the final electron
        acceptor in electron transport
        phosphorylation.
         a. hydrogen
         b. carbon
    *   c. oxygen
         d. H$_2$O
         e. NAD$^+$

M 91. The correct operational sequence of the three
        processes listed below is:
         a. glycolysis >> oxidative
             phosphorylation >> Krebs
         b. oxidative phosphorylation >>
             glycolysis >> Krebs
         c. Krebs >> glycolysis >> oxidative
             phosphorylation
         d. oxidative phosphorylation >> Krebs
             >> glycolysis
    *   e. glycolysis >> Krebs >> oxidative
             phosphorylation

M 92. The greatest number of ATP molecules is
        produced in
         a. glycolysis.
         b. lactate fermentation.
         c. anaerobic respiration
    *   d. electron transport and oxidative
             phosphorylation.
         e. the Krebs cycle.

E 93. When glucose is used as the energy source,
        the largest amount of ATP is produced in
         a. glycolysis.
         b. acetyl CoA formation.
         c. the Krebs cycle.
         d. substrate-level phosphorylation.
    *   e. electron transport and oxidative
             phosphorylation.

M   94. What is the name of the process by which reduced NAD^+ transfers electrons to oxygen?
a. glycolysis
b. acetyl CoA formation
c. the Krebs cycle
* d. electron transport and oxidative phosphorylation
e. substrate-level phosphorylation

E   95. During electron transport phosphorylation, which ions accumulate in the outer compartment of the mitochondria?
a. calcium
* b. hydrogen
c. oxygen
d. phosphorus
e. sodium

M   96. The ultimate electron acceptor in aerobic respiration is
a. NAD^+
b. CO_2
c. ADP
d. $NADP^+$
* e. O_2

M   97. The energy used to generate most of the ATP formed in aerobic respiration is released when electrons ultimately are passed from NADH to which of the following?
* a. oxygen
b. acetyl CoA
c. FADH
d. CO_2
e. NADPH

E   98. The total yield of ATP from one molecule of glucose is
a. 2.
b. 8.
* c. 32.
d. 36.
e. 48.

E   99. Under anaerobic conditions muscle cells produce
a. ethyl alcohol.
b. acetaldehyde.
c. pyruvate.
* d. lactate.
e. citrate.

D   100. Fermentation
* a. may occur in a muscle under anaerobic conditions.
b. produces more ATP than is liberated in the hydrogen transfer series.
c. breaks down glucose in reaction with oxygen.
d. is restricted to yeasts.
e. none of these

ALTERNATIVE ENERGY SOURCES IN THE HUMAN BODY

M   101. When blood glucose levels decrease (as between meals), what reserves are tapped?
* a. glycogen
b. fats
c. proteins
d. steroids
e. amino acids

M   102. When proteins and fats are used as energy sources, their breakdown subunits enter
a. glycolysis
b. electron transport
* c. Krebs cycle
d. chemiosmosis
e. fermentation

Matching Questions

M **103.** Choose the one most appropriate answer for each.

1 ____ microtubules
2 ____ peroxisomes
3 ____ Golgi bodies
4 ____ DNA molecules
5 ____ RNA molecules
6 ____ nucleus
7 ____ lysosomes
8 ____ mitochondria
9 ____ nucleoli
10 ____ ribosomes

A. contain enzymes for intracellular digestion

B. primary cellular organelles where proteins are assembled

C. package cellular secretions for export

D. extract energy stored in carbohydrates; synthesize ATP; produce water and CO2

E. synthesize subunits that will be assembled into two-part ribosomes in the cytoplasm

F. transcription, translation of hereditary instructions into specific proteins

G. physical home of chromosomes

H. encoding hereditary information

I. help distribute chromosomes to the new cells during cell division

J. convert hydrogen peroxide to harmless substances

Answers:
1. I	2. J	3. C
4. H	5. F	6. G
7. A	8. D	9. E
10. B		

M **104.** Matching. Choose the one most appropriate answer for each.

1 ____ glycolysis
2 ____ fermentation
3 ____ acetyl-CoA formation
4 ____ the Krebs cycle
5 ____ electron transport phosphorylation

A. produces NADH and CO_2; changes pyruvate

B. produces ATP, NADH, and CO_2

C. splits glucose into two pyruvate molecules

D. regenerates NAD^+ as pyruvate is converted to lactate

E. uses a membrane-bound system that contains cytochromes to produce ATP

Answers:
1. C	2. D	3. A
4. B	5. E	

Classification Questions

The following items a–e are organelles found in animal cells. Answer questions 105–113 with reference to these organelles.

a. ribosomes
b. mitochondria
c. lysosomes
d. Golgi bodies
e. endoplasmic reticulum

E **105.** These are the structures upon which proteins are assembled.

M **106.** The cellular digestion and disposal of biological molecules occurs inside these organelles.

M **107.** Aerobic respiration occurs in these organelles.

M **108.** RNA carries out the translation of the genetic code in association with ribosomes on this organelle.

M **109.** The packaging of secretory proteins occurs in association with these structures.

M **110.** This organelle is involved in lipid production and protein transport.

D   **111.** The hemoglobin of mammals and birds is synthesized on these tiny, two-part organelles.

E   **112.** Sugar metabolism occurs in association with this organelle.

D   **113.** DNA synthesis occurs in the nucleus. Its breakdown can occur in this organelle.

Answers: 105. a   106. c   107. b
                108. e   109. d   110. e
                111. a   112. b   113. c

Questions 114 –118 ask about membrane permeability. Answer them in reference to the five processes below:

       a. simple diffusion
       b. facilitated diffusion
       c. osmosis
       d. active transport
       e. endocytosis

E   **114.** This process would be used by white blood cells to ingest bacteria.

E   **115.** This process specifically moves water molecules across a differentially permeable membrane.

E   **116.** This process explains the movement of any kind of molecule from areas of higher concentration to ones of lower concentration.

E   **117.** This is the process whereby a protein assists in simple diffusion.

M   **118.** This explains the movement of molecules against a concentration gradient.

Answers: 114. e   115. c   116. a
                117. b   118. d

Use the five processes listed below to answer questions 119-123.

       a. glycolysis
       b. aerobic respiration
       c. anaerobic electron transport
       d. Krebs
       e. lactate fermentation

E   **119.** In this process two ATP, eight NADH, and carbon dioxide are produced.

E   **120.** In this process the final product is lactate.

M   **121.** This process yields the most energy.

D   **122.** This process involves electron transport (oxidative) phosphorylation

E   **123.** This process precedes the Krebs cycle.

Answers: 119. d   120. e   121. b
                122. b   123. a

Use the five compounds listed below to answer questions 124-128.

       a. ethanol
       b. pyruvate
       c. lactate
       d. citrate
       e. glucose

M   **124.** This compound is the immediate substrate in lactate fermentation.

M   **125.** This compound is the most likely end product of a human runner experiencing an oxygen debt.

D   **126.** This compound is the initial substrate on which on cellular respiration proceeds.

M   **127.** This compound is the end product of glycolysis.

E   **128.** This compound is an end product of anaerobic respiration in exercising muscle.

Answers: 124. b   125. c   126. e
                127. b   128. c

Selecting the Exception

M   129.   Four of the five answers listed below are familiar organelles in the cytoplasm. Select the exception.
      *   a.   nucleolus
          b.   mitochondria
          c.   ribosome
          d.   Golgi apparatus
          e.   chloroplast

M   130.   Four of the five answers listed below are features of plasma membrane extensions. Select the exception.
      *   a.   amyloplast
          b.   centriole
          c.   microtubule
          d.   basal body
          e.   9+2 array

M   131.   Four of the five answers listed below are bound by membranes. Select the exception.
          a.   mitochondria
      *   b.   ribosome
          c.   peroxisome
          d.   nucleus
          e.   lysosome

M   132.   Four of the five answers listed below are characteristics of the plasma membrane. Select the exception.
          a.   phospholipid
          b.   fluid mosaic
          c.   lipid bilayer
      *   d.   inert and impermeable
          e.   hydrophobic tails

D   133.   Three of the four answers listed below are related by energy requirements. Select the exception.
          a.   active transport
          b.   endocytosis
      *   c.   facilitated diffusion
          d.   exocytosis

M   134.   Four of the five answers listed below are related by their description of enzyme properties. Select the exception.
          a.   cofactors
          b.   active sites
          c.   reusable
      *   d.   substrate
          e.   catalyst

D   135.   Four of the five answers listed below are cofactors or coenzymes. Select the exception.
          a.   mineral
          b.   water soluble vitamin
          c.   metallic ions
          d. NAD^+
      *   e.   protein

D   136.   Three of the four answers listed below are parts of a common molecule. Select the exception.
          a.   phosphate group
          b.   adenine
      *   c.   deoxyribose
          d.   ribose

D   137.   Four of the five answers listed below are hydrogen acceptors. Select the exception.
          a.   oxygen
          b.   cytochrome
      *   c.   ATP
          d. NAD^+
          e.   FAD

D   138.   Four of the five answers listed below are compounds associated with anaerobic respiration. Select the exception.
          a.   pyruvate
          b.   lactic acid
          c.   glucose
      *   d.   oxaloacetate
          e.   phosphoglyceraldehyde

CHAPTER 4

INTRODUCTION TO TISSUES, ORGAN SYSTEMS, AND HOMEOSTASIS

Multiple-Choice Questions

THE BODY IN BALANCE

E 1. Chemical and structural bridges link groups or layers of similar cells, uniting them in structure and function as
   a. organs.
   b. organ systems.
  * c. tissues.
   d. cuticles.

E 2. Different tissues integrated and functioning with a common purpose defines a
   a. organisms.
   b. organ systems.
   c. subtissues.
  * d. organs

E 3. Which of the following is NOT one of the basic types of tissues in the human body?
   a. connective
  * b. ectoderm
   c. nervous
   d. epithelial
   e. muscle

E 4. Stomach, spleen, liver, and pancreas are examples of
   a. organ systems.
   b. tissues.
  * c. organs.
   d. cells.

E 5. Circulatory, digestive, reproductive, and excretory are all examples of
  * a. systems.
   b. organs.
   c. tissues.
   d. cells.

EPITHELIAL TISSUE

E 6. The tissue that lines internal surfaces of the body is
  * a. epithelial.
   b. loose connective.
   c. supportive connective.
   d. fibrous.
   e. adipose.

M 7. Epithelial cells are specialized for all the following functions EXCEPT
   a. secretion.
   b. protection.
   c. filtration.
  * d. contraction.
   e. absorption.

E 8. Tears, milk, sweat, and oil are secreted by glands made of what tissue?
  * a. epithelial
   b. loose connective
   c. lymphoid
   d. nervous
   e. adipose

E 9. What type of tissue is characterized by adherence to a basement membrane on one side and a free surface on the opposite?
   a. connective
   b. muscle
   c. nervous
  * d. epithelial
   e. adipose

E 10. The principal difference between simple and stratified epithelium is in the number of
  * a. cell layers.
   b. total cells.
   c. nuclei per cell.
   d. free surfaces.
   e. microvilli.

M 11. What type of epithelium would you expect to find where there is considerable secretion of enzymes such as the small intestine?
   a. cuboidal
   b. squamous
  * c. columnar
   d. pseudostratified
   e. basement

M   12. Which of the following statements about exocrine glands in NOT accurate?
     a. Products may be secreted onto a free surface.
     b. Secretions reach their destination by way of tubes and ducts.
  * c. Hormones are produced by exocrines.
     d. They can be classified as simple or compound.
     e. Apocrine is a type of exocrine gland.

E   13. Sebaceous glands
     a. secrete oil.
     b. are holocrine.
     c. release entire cells with the secretion.
     d. are exocrine glands.
  * e. all of these.

E   14. Exocrine glands secrete
     a. enzymes.
     b. sweat.
     c. milk.
     d. saliva.
  * e. all of these

M   15. Which epithelial cell is modified for diffusion?
     a. cuboidal
  * b. simple squamous
     c. simple columnar
     d. stratified squamous
     e. stratified columnar

M   16. The type of epithelial cell found in the lining of the stomach, intestinal tract, and part of the respiratory tract is
     a. simple cuboidal.
     b. simple squamous.
  * c. simple columnar.
     d. stratified.
     e. stratified columnar.

CONNECTIVE TISSUES

M   17. Which of the following is NOT included in connective tissues?
     a. bone
  * b. skeletal muscle
     c. cartilage
     d. collagen
     e. blood

E   18. What type of tissue is blood?
     a. epithelial
     b. muscular
  * c. connective
     d. adipose
     e. noncellular fluid

M   19. An extracellular ground substance is characteristic of
     a. muscle tissue.
     b. epithelial tissue.
  * c. connective tissue.
     d. nervous tissue.
     e. embryonic tissue.

M   20. Connective tissues include all the following EXCEPT
     a. cartilage.
     b. blood.
     c. bone.
     d. fat.
  * e. outer skin.

E   21. Dense, regular tissues that connect bone to bone are called
     a. muscles.
     b. cartilage.
  * c. ligaments.
     d. tendons.
     e. all of these

M   22. Collagen fibers are characteristic of which tissue?
     a. muscle
     b. epithelial
  * c. connective
     d. nervous
     e. embryonic

M   23. Tendons connect
     a. bones to bones.
     b. bones to ligaments.
  * c. muscles to bones.
     d. bones to cartilage.
     e. all of these

E   24. Cartilage is found
     a. in the nose.
     b. at the ends of bones.
     c. in the external ear.
     d. between vertebrae.
  * e. all of these

M **25.** What is the name given to the type of cartilage that is found in the disks between the vertebrae?
   a. chondrocytes
   b. elastic
   c. compact
   d. hyaline
   * e. fibrocartilage

M **26.** All of the following could be used to describe some aspect of a cartilage cell EXCEPT
   a. chondrocyte.
   * b. osteocyte.
   c. lacunae.
   d. chondroblast.

E **27.** Marrow is found
   a. in compact bone.
   b. embedded in subcutaneous adipose.
   * c. in spongy bone.
   d. in blood.
   e. between the dermis and epidermis.

E **28.** Bone cells are called
   a. chondrocytes.
   b. macrophages.
   c. lacunae.
   * d. osteocytes.
   e. fibroblasts.

M **29.** Which of the following is NOT a part of the "formed elements" of the blood?
   a. red blood cells
   * b. plasma
   c. white blood cells
   d. platelets
   e. megakaryocytes

E **30.** The fluid portion of the blood is called
   * a. plasma.
   b. formed elements.
   c. interstitial.
   d. extracellular matrix.

E **31.** Adipose tissue cells are filled with
   a. minerals.
   * b. fat.
   c. cartilage.
   d. fibers.
   e. muscles.

MUSCLE TISSUE

M **32.** If its cells are striated and fused at the ends so that the cells contract as a unit, the tissue is
   a. smooth muscle.
   b. dense fibrous connective.
   c. supportive connective.
   * d. cardiac muscle.
   e. none of these

E **33.** Muscle that is NOT striped and is involuntary is
   a. cardiac.
   b. skeletal.
   c. striated.
   * d. smooth.
   e. both cardiac and smooth

E **34.** Cardiac muscle cells are
   a. involuntary.
   b. voluntary.
   c. striated.
   d. slow contracting.
   * e. both involuntary and striated

E **35.** Smooth muscles are
   a. striated and voluntary.
   b. isolated, spindle-shaped cells.
   c. found in the walls of hollow structures such as blood vessels and the stomach.
   d. involuntary and nonstriated.
   * e. all except striated and voluntary

NERVOUS TISSUE

E **36.** The basic cells that the body uses for rapid communication and control are
   * a. neurons.
   b. neuroglia.
   c. chondrocytes.
   d. formed elements.
   e. nerves.

E **37.** Accessory cells in nervous tissue are called
   a. neurons.
   * b. neuroglia.
   c. chondrocytes.
   d. formed elements.
   e. nerves.

E   38. Outgoing messages are conducted by
         a. an axon.
         b. any process that is long enough to
            reach the receptor.
    *   c. numerous dendrites.
         d. the intervention of the neuroglia.
         e. chondrocytes.

M   39. Clusters of neuronal processes are called
         a. dendrites.
         b. chondrocytes.
         c. neuroglia.
    *   d. nerves.
         e. axons.

M   40. Receptors transmit to the central nervous
         system by way of
    *   a. sensory neurons.
         b. neuroglia.
         c. dendrites only.
         d. motor neurons.
         e. axons only.

E   41. Impulses are received by nerve cells through
         a. an axon.
         b. any process that is long enough to
            reach the receptor.
    *   c. numerous dendrites.
         d. the intervention of the neuroglia.
         e. chondrocytes.

CELL-TO-CELL CONTACTS

M   42. What type of junction would be found in an
         organ such as the stomach which is
         subjected to considerable stretching?
         a. gap junction
    *   b. desmosome
         c. tight junction
         d. microvillus
         e. merocrine

M   43. What would be used between cells to keep
         molecules from freely crossing the
         epithelium?
         a. gap junction
         b. desmosome
    *   c. tight junction
         d. microvillus
         e. merocrine

M   44. What would form open channels that
         directly link the cytoplasms of adjacent
         cells?
    *   a. gap junction
         b. desmosome
         c. tight junction
         d. microvillus
         e. merocrine

MEMBRANES

M   45. Membranes that do NOT contain gland cells
         are termed
         a. mucous.
         b. dense, regular.
         c. desmosomes.
         d. apocrine.
    *   e. serous.

M   46. Membranes that line the tubes and cavities
         of the digestive, respiratory, and
         reproductive systems are called
    *   a. mucous.
         b. dense, regular.
         c. desmosomes.
         d. apocrine.
         e. serous.

ORGAN SYSTEMS

E   47. The endocrine system functions in
         a. conduction.
         b. contraction.
    *   c. hormonal control of body functioning.
         d. protection against disease.
         e. cell production.

E   48. The maintenance of the volume and
         composition of body fluids is the direct
         responsibility of which system?
         a. integumentary
         b. immune
         c. digestive
    *   d. urinary
         e. circulatory

M   49. Which system is involved with heat
         production?
         a. endocrine system
         b. nervous system
    *   c. muscular system
         d. respiratory system
         e. skeletal system

E   50. Integration of body functions is controlled
         by the
         a. respiratory system.
         b. nervous system.
         c. endocrine system.
         d. defense system.
       * e. both the nervous and endocrine
            systems.

M   51. Which system produces blood cells?
         a. endocrine
       * b. skeletal
         c. muscular
         d. defense
         e. integumentary

E   52. The central nervous system is housed in the
       * a. cranial and spinal cavities.
         b. cranial cavity only.
         c. thoracic cavity.
         d. pelvic cavity.
         e. abdominal cavity.

E   53. The thoracic cavity contains the heart and
         a. stomach.
       * b. lungs.
         c. liver.
         d. brain.
         e. pancreas.

SKIN: EXAMPLE OF AN ORGAN SYSTEM

E   54. The word *integument* is derived from the
         word for
         a. protection.
         b. support.
       * c. covering.
         d. contraction.
         e. resistance.

E   55. Melanin protects the skin from
         a. desiccation.
         b. abrasion.
       * c. ultraviolet radiation.
         d. infrared damage.
         e. invasion by bacteria.

E   56. The integumentary system is responsible
         for all but which of the following?
         a. protection against bacterial attack
         b. protection against abrasion
         c. synthesis of certain vitamins
       * d. blood cell formation
         e. control of temperature and prevention
            of drying out

E   57. The largest organ of the vertebrate body is
         which of the following?
         a. lungs
         b. liver
         c. stomach
       * d. skin
         e. small intestines

E   58. Vitamin D is required for _____
         metabolism.
         a. sulfur
         b. phosphorus
       * c. calcium
         d. potassium
         e. zinc

M   59. Which of the following is NOT found in
         the epidermis?
         a. stratified epithelium
       * b. blood vessels
         c. tight cell junctions
         d. keratin
         e. melanin

E   60. Which of the following statements is false
         concerning the outermost layer of the
         epidermis?
         a. It is the first to experience any
            abrasion.
         b. Keratin provides waterproofing.
         c. Millions of cells are worn off daily.
       * d. Its cells are undergoing rapid cell
            division.
         e. It is called the stratum corneum.

HOMEOSTASIS AND SYSTEMS CONTROL

M   61. Which are examples of integrators?
       * a. brain, spinal cord
         b. muscles, glands
         c. sensory cells in eye, tongue, and ear
         d. bones

E   62. The control of the temperature of the body
         is an example of which of the following?
         a. homeostatic mechanism
         b. positive feedback system
         c. endocrine function
         d. negative feedback system
       * e. both homeostatic and negative feedback
            systems.

M **63.** Which is the correct sequence involved in the regulation of organ systems?
- a. stimulus, receptor, integrator, response, effector
- b. stimulus, response, integrator, receptor, effector
- * c. stimulus, receptor, integrator, effector, response
- d. stimulus, integrator, receptor, effector, response
- e. stimulus, effector, integrator, receptor, response

M **64.** An effector can be a
- a. muscle.
- b. nerve.
- c. gland.
- d. receptor.
- * e. both muscle and gland.

M **65.** Which involves a positive feedback stimulation?
- a. temperature control
- * b. sexual stimulation
- c. glucose concentration
- d. absorption of toxins
- e. muscle concentration

Matching Questions

D **66.** Choose the one most appropriate answer for each.

1 ____ adipose tissue
2 ____ blood
3 ____ dense connective tissue
4 ____ glandular epithelium
5 ____ loose connective tissue
6 ____ interstitial fluid
7 ____ neuron
8 ____ epidermis

A. tendons are made of this

B. contains collagen and elastin; acts as a packing material that supports internal organs

C. receives, conducts, and initiates signals in response to environmental changes

D. stores fatty reserves

E. offers resistance to mechanical injury and loss of internal fluids; also a barrier against microorganisms

F. secretes extracellular products such as sweat, mucus, tears, and shells

G. fluid ground substance plus free cells; involved in transport, pH, and temperature stability

H. extracellular fluid that bathes cells and tissues

Answers:
1. D	2. G	3. A.
4. F	5. B	6. H
7. C	8. E	

Classification Questions

Answer questions 67–71 in reference to the five types of connective tissue listed below:

    a. loose tissue
    b. dense regular tissue
    c. adipose
    d. cartilage
    e. blood

E   **67.** Tendons are composed of this tissue.

E   **68.** The elasticity of skin is due to the presence of this kind of tissue.

M   **69.** This tissue plays an important role in stabilizing human body temperature.

E   **70.** This tissue provides nourishment to each of the other connective tissues.

M   **71.** This tissue plays a role in assuring a long-term energy reserve.

Answers: 67. b   68. a   69. e
                70. e   71. c

Answer questions 72–76 in reference to the five organ systems listed below:

    a. circulatory
    b. lymphatic
    c. digestive
    d. endocrine
    e. respiratory

E   **72.** The absorption of nutrients into the blood is the responsibility of this system.

E   **73.** This system returns tissue fluid to the blood.

M   **74.** Airborne allergens are first encountered by this system.

E   **75.** This system rapidly transports many vital materials throughout the body.

E   **76.** The production of hormones occurs in this system.

Answers: 72. c   73. b   74. e
                75. a   76. d

Selecting the Exception

M   **77.** Four of the five answers listed below are secreted by an exocrine gland. Select the exception.
    a. wax
    b. saliva
   * c. hormone
    d. milk
    e. mucus

D   **78.** Four of the five answers listed below are related by a common tissue type. Select the exception.
    a. adipose
    b. bone
    c. cartilage
    d. blood
   * e. epithelium

M   **79.** Four of the five answers listed below are functions of the skeleton. Select the exception.
   * a. controls body temperature
    b. produces blood cells
    c. protection
    d. storage sites for calcium and phosphorus
    e. muscle attachment

CHAPTER 5
THE MUSCULAR AND SKELETAL SYSTEMS

Multiple-Choice Questions

CHARACTERISTICS OF BONE

M   **1.** Which of the following would NOT be a function of bone?
   a. movement of the body
   b. support
   *   c. hormone secretion
   d. mineral storage
   e. blood cell formation

E   **2.** Mature, living bone cells are called
   a. osteoblasts.
   *   b. osteocytes.
   c. osteolytes.
   d. osteoclasts.
   e. osteopores.

M   **3.** Lacunae are
   a. spaces inside a bone cell.
   b. the places where marrow is located.
   c. the location of the contractile units of muscle.
   *   d. spaces within the bone ground substance.
   e. found only within cartilage.

E   **4.** Which of the following would NOT be found inside a Haversian canal?
   *   a. bone cell
   b. blood capillary
   c. nerve
   d. nutrients
   e. wastes

D   **5.** Which of the following choices is out of place with the rest?
   *   a. yellow marrow
   b. red marrow
   c. spongy bone
   d. blood cells
   e. latticework

E   **6.** Where would expect to find the epiphyses?
   a. shaft of a bone
   b. marrow cavity of bone
   c. origin of a muscle
   *   d. ends of long bones
   e. sutures of the skull

E   **7.** In spongy bone tissue the spaces are filled with
   a. air.
   b. blood.
   c. cartilage.
   *   d. marrow.
   e. lymph.

M   **8.** Haversian canals are characteristic of which tissue?
   a. adipose
   *   b. bone
   c. cartilage
   d. epithelial
   e. muscular

D   **9.** All but which of the following are associated with bone formation?
   a. osteoblasts
   b. cartilage
   *   c. osteoporosis
   d. marrow cavity formation
   e. calcium

M   **10.** Growth of long bones
   a. follows the cartilage model.
   b. occurs in the middle at first, then at both ends.
   c. is characterized by bone tissue replacing calcified cartilage.
   d. is characterized by the persistence of cartilage at both ends of the shaft.
   *   e. all of these

D   **11.** If some bleached bones found lying in the desert were carefully examined, which of the following would NOT be present?
   *   a. osteocytes
   b. Haversian canals
   c. calcium
   d. marrow cavity
   e. compact bone

HOW THE SKELETON GROWS AND IS MAINTAINED

E   **12.** Osteoporosis is a condition characterized by
- a. excessive build up of bone calcium deposits.
- b. hardening of the arteries in the bone.
- *   c. loss of bone mass.
- d. excessive rigidity of some bones.
- e. curvature of the spine.

M   **13.** Which statement is false?
- a. Calcium is the most important mineral involved with bone tissue turnover.
- b. Osteoblasts and osteoclasts are involved with the reabsorption and repair of bones.
- c. Bone mass decreases with age.
- *   d. Males have greater problems with loss of bone tissue than females do.
- e. Marrow fills the cavities in the spaces of the spongy bone and in the center of the shaft of the bone almost as quickly as the spaces are formed.

M   **14.** Tendons connect
- a. bones to bones.
- b. bones to ligaments.
- *   c. muscles to bones.
- d. bones to cartilage.
- e. all of these

M   **15.** Ligaments connect
- *   a. bones to bones.
- b. bones to ligaments.
- c. muscles to bones.
- d. bones to cartilage.
- e. all of these

M   **16.** The hormones associated with bone remodeling are
- a. glucagon and insulin.
- b. insulin and calcitonin.
- c. growth hormone and insulin.
- *   d. calcitonin and parathyroid hormone.
- e. prolactin and thyroxine.

THE AXIAL SKELETON

M   **17.** The human axial skeleton includes all of the following EXCEPT
- a. skull.
- b. ribs.
- *   c. pectoral girdle.
- d. sternum.
- e. vertebral column.

M   **18.** Which of the following bones is NOT a part of the human skull?
- *   a. iliac
- b. frontal
- c. ethmoid
- d. occipital
- e. temporal

M   **19.** The foramen magnum is a passageway for
- a. nasal cavities.
- b. eye sockets.
- *   c. spinal cord.
- d. Haversian canals.
- e. synovial fluid.

M   **20.** Which of the following are associated with the nasal cavity?
- a. palatine bones
- b. zygomatic bones
- c. vomer bone
- d. palatine bones and zygomatic bones
- *   e. palatine bones and vomer bone

THE APPENDICULAR SKELETON

E   **21.** Which of the following is NOT part of the appendicular skeleton?
- a. clavicle
- b. scapula
- c. fibula
- *   d. ribs
- e. patella

E   **22.** The bone in the upper arm is the
- a. radius.
- b. ulna.
- c. tibia.
- *   d. humerus.
- e. femur.

E   **23.** Bones in fingers or toes are called
- a. hyoid.
- b. patella.
- c. scapula.
- d. clavicle.
- *   e. phalanges.

E   **24.** The kneecap is the
       a. hyoid.
    *   b. patella.
       c. scapula.
       d. clavicle.
       e. phalanx.

E   **25.** The collarbone is the
       a. hyoid.
       b. patella.
       c. scapula.
    *   d. clavicle.
       e. phalanx.

E   **26.** The shoulder blade is the
       a. hyoid.
       b. patella.
    *   c. scapula.
       d. clavicle.
       e. phalanx.

E   **27.** Bones such as the humerus and femur are examples of which kind of bones?
    *   a. long
       b. short
       c. flat
       d. irregular
       e. all of these

M   **28.** The bone of the upper arm is connected to the bones of the lower arm by
       a. tendons.
       b. lordosis.
       c. phalanges.
    *   d. ligaments.
       e. marrow.

E   **29.** The quantity 12 applies to which of these vertebral groupings?
       a. lumbar
       b. cervical
       c. coccyx
    *   d. thoracic
       e. sacrum

E   **30.** The ribs are attached on the ventral (front) of the human body to the
       a. vertebrae.
    *   b. sternum.
       c. intervertebral disks.
       d. sacrum.
       e. coccyx

E   **31.** Which of the following are found in BOTH the pectoral and pelvic girdles?
       a. tarsals
    *   b. phalanges
       c. carpals
       d. coxals
       e. metacarpals

E   **32.** Which of the following bones in the lower leg has no counterpart in the lower arm?
       a. tibia
       b. fibula
    *   c. patella
       d. tarsals
       e. metatarsals

JOINTS

M   **33.** The nonmoving joints between skull bones are examples of what kind of joints?
       a. synovial
    *   b. fibrous
       c. cartilaginous
       d. hinge
       e. none of these

M   **34.** The vertebral discs with small amounts of movement are examples of what kind of joints?
       a. synovial
       b. fibrous
    *   c. cartilaginous
       d. hinge
       e. none of these

D   **35.** Which of the following would NOT be considered a synovial joint?
       a. hip bone to femur
       b. elbow
    *   c. joints between vertebrae
       d. shoulder to upper arm
       e. knee

M   **36.** Which of the following types of joints allows the LEAST amount of movement?
    *   a. fibrous
       b. ball and socket
       c. hingelike
       d. synovial
       e. cartilaginous

THE MUSCULAR SYSTEM AND HOW IT INTERACTS WITH THE SKELETON

E 37. The human biceps and triceps muscles may be described best as working
      a. independently.
      b. synergistically.
  * c. antagonistically.
      d. synchronistically.
      e. involuntarily.

M 38. The biceps muscle extends from the shoulder to the lower arm which is raises. The __?__ is on the lower arm.
      a. origin
      b. elevator
      c. fulcrum
  * d. insertion
      e. flexor

E 39. Smooth muscle is
  * a. involuntary and nonstriated.
      b. responsible for movement of the skeleton.
      c. involved in contraction of the heart.
      d. connected to bones by tendons.
      e. both involuntary and nonstriated, plus is involved in contraction of the heart.

M 40. The ability to extend a leg is the result of
      a. contraction of ligaments and tendons.
  * b. contraction of a muscle.
      c. lengthening of a muscle.
      d. a combination of push and pull by antagonistic muscle pairs.

E 41. Each muscle fiber is also called a
  * a. muscle.
      b. muscle cell.
      c. myofibril.
      d. sarcomere.
      e. all of these

M 42. The gastrocnemius muscle is located
      a. in the forelimb.
      b. on the back.
      c. in the hip area.
  * d. in the lower leg.
      e. in the neck.

E 43. The pectoralis major muscle is located
  * a. in the chest.
      b. on the back.
      c. near the hips.
      d. in the upper leg.
      e. in the lower leg.

D 44. Reciprocal innervation of reflexes between antagonistic muscle pairs
  * a. is the usual basis of coordinated contractions.
      b. is the means by which rods prevent cones from being stimulated.
      c. refers to the lens adjustments that bring about precise focusing onto the retina.
      d. explains the mechanism for the operation of the calcium pump.
      e. all of these

A CLOSER LOOK AT MUSCLES

M 45. Which of the following includes all the others?
      a. actin
      b. myofibril
      c. myosin
      d. myofilament
  * e. muscle cell

M 46. During muscle contractions
      a. the myofibrils shorten.
      b. the actin and myosin filaments slide over each other.
      c. the actin filaments move toward the middle of the sacromere during contraction and away on relaxation.
      d. the muscle thickens.
  * e. all of these

D 47. During contraction
      a. cross bridges of muscle filaments are broken and reformed.
      b. ATP is used to form cross-bridges.
      c. muscle cells use glycogen as their energy source.
      d. if there is a poor supply of oxygen, glycogen depletion by glycolysis will lead to fatigue.
  * e. all of these

D 48. In their action, muscles would be most like
  * a. ropes.
      b. levers.
      c. push rods.
      d. screws.
      e. hammers.

M 49. In rigor mortis, the muscles of the body
    lose their ability to
       a. contract.
       b. flex.
   *   c. relax.
       d. form cross-bridges.
       e. bend.

CONTROL OF MUSCLE CONTRACTION

M 50. The element specifically associated with
    muscle contraction is
       a. phosphorus.
       b. potassium.
   *   c. calcium.
       d. sodium.
       e. chlorine.

E 51. The most immediate, but necessarily
    limited, source of energy for reformation of
    ATP in muscle cells is
       a. aerobic respiration.
       b. mitochondrial pathways.
       c. electron transport phosphorylation.
   *   d. creatine phosphate.
       e. anaerobic fermentation.

M 52. Which substance is released by motor
    neurons to initiate a muscle contraction?
   *   a. acetylcholine
       b. dopamine
       c. serotonin
       d. noradrenalin
       e. all of these

E 53. Functionally, the plasma membrane of a
    muscle cell is most like that of a
       a. bone cell.
   *   b. nerve cell.
       c. cartilage cell.
       d. pancreatic islet cell.
       e. epidermal cell.

M 54. Calcium ions in muscle cells are stored in
       a. vacuoles.
   *   b. sarcoplasmic reticulum.
       c. mitochondria.
       d. cross-bridges.
       e. acetylcholine.

D 55. One of the functional proteins constituting
    the actin filaments is
       a. myosin.
       b. keratin.
       c. elastin.
   *   d. tropomyosin.
       e. epithelium.

PROPERTIES OF WHOLE MUSCLES

M 56. Which of the following is NOT true of
    "fast" muscle in humans?
       a. rapid contractions
       b. powerful
       c. fewer blood capillaries
   *   d. abundant myoglobin
       e. fewer mitochondria

M 57. A motor neuron and all the muscles under
    its control is called what kind of unit?
       a. end
       b. movement
       c. muscle
   *   d. motor
       e. coordination

M 58. An active, nonfatiguing muscle would be
    expected to have
       a. aerobic respiration.
       b. numerous mitochondria.
       c. moderate rates of contraction.
       d. aerobic respiration and moderate rates of
          contraction.
   *   e. aerobic respiration, numerous
          mitochondria, and moderate rates of
          contraction.

E 59. Muscle fatigue is a result of
   *   a. accumulation of lactic acid.
       b. exhaustion of available ATP.
       c. reduction in lactic acid and oxygen debt.
       d. failure of calcium channels to open
          after prolonged use.

D 60. Which of the following statements does
    NOT describe normal activity in a motor
    system?
       a. All motor systems require the presence
          of some medium or structural element
          against which force can be applied.
       b. In a skeletal muscle system,
          coordinated contraction depends on
          reciprocal innervation of motor neurons
          to antagonistic muscle pairs.
       c. In vertebrates, only skeletal muscle acts
          to move the body through the
          environment.
   *   d. In a resting muscle, energy is stored in
          the form of tropomyosin.
       e. all of these

M   **61.** Which of the following statements is false?
  a. Calcium ions are released as an action potential is propagated along a skeletal muscle cell.
  * b. Acetylcholine provides energy to propel actin filaments past myosin filaments during muscle cell activity.
  c. Calcium ions are actively collected and stored in the sarcoplasmic reticulum as the muscle returns to the resting state.
  d. The interaction of acetylcholine with receptors on the sarcolemma may produce graded potentials.
  e. Tetanus is the normal mode of contraction in human skeletal muscle.

D   **62.** Which of the following statements is true of the all-or-none principle?
  a. It explains muscle fatigue.
  * b. It describes the contraction of individual muscle cells.
  c. A muscle, such as the biceps, either contracts or it doesn't.
  d. Muscle tone is the direct result of this phenomenon.
  e. It is equivalent to tetany.

Classification Questions

Answer questions 63-67 in reference to the five bones listed below:

  a. clavicle
  b. lumbar vertebrae
  c. tibia
  d. metatarsal
  e. metacarpal

E   **63.** All of the above are part of the appendicular skeleton EXCEPT

M   **64.** This bone is extremely subject to breakage in contact sports activity.

M   **65.** If one had a slipped disk, that disk might be next to this bone.

M   **66.** This bone is an ankle bone in humans.

M   **67.** This bone connects to the femur.

Answers: 63. b   64. a   65. b
               66. d   67. c

Answer questions 68-72, in reference to the five muscles listed below:

  a. pectoralis major
  b. deltoid
  c. rectus femoris
  d. triceps
  e. gastrocnemius

D   **68.** Which muscle is a principal muscle of the upper leg?

M   **69.** Which muscle is located on the upper shoulder?

M   **70.** Which muscle is antagonistic to the action of the biceps?

D   **71.** The Achilles tendon attaches which muscle to the heel bones?

D   **72.** Which muscle is the principal muscle extending across the chest?

Answers: 68. c   69. b   70. d
               71. e   72. a

Selecting the Exception

M   **73.** Four of the five answers listed below are parts of the same anatomical area. Select the exception.
  a. humerus
  * b. fibula
  c. radius
  d. clavicle
  e. scapula

M   **74.** Four of the five answers listed below are parts of the same skeletal division. Select the exception.
  a. cranium
  b. ribs
  c. sternum
  d. vertebrae
  * e. phalanges

M   **75.** Four of the five answers listed below are regions of the vertebral column. Select the exception.
  a. cervical
  * b. appendicular
  c. lumbar
  d. thoracic
  e. sacral

E   **76.** Four of the five answers listed below are
         types of bones. Select the exception.
    *   a. immovable
         b. long
         c. short
         d. flat
         e. irregular

E   **77.** Four of the five answers listed below are
         muscles. Select the exception.
         a. pectoralis
    *   b. patella
         c. gastrocnemius
         d. deltoid
         e. sartorius

D   **78.** Four of the five answers listed below are
         molecules that participate in muscle
         contraction. Select the exception.
    *   a. sarcolemma
         b. ATP
         c. calcium
         d. actin
         e. myosin

CHAPTER 6

DIGESTION AND NUTRITION

Multiple-Choice Questions

OVERVIEW OF THE DIGESTIVE SYSTEM

E **1.** The lumen of the digestive tract describes
   a. the movement of food along its length
   b. its outer covering.
   * c. the space inside.
   d. muscular layers.
   e. twisted path it follows through the abdomen.

M **2.** Which of the following is NOT an accessory organ to the digestive system?
   a. pancreas
   * b. spleen
   c. salivary glands
   d. liver
   e. gallbladder

M **3.** The process that moves nutrients into the blood or lymph is
   * a. absorption.
   b. assimilation.
   c. digestion.
   d. ingestion.
   e. elimination.

D **4.** The process that releases digestive enzymes is
   a. absorption.
   b. assimilation.
   * c. secretion.
   d. digestion.
   e. elimination.

M **5.** Which of the following layers of the digestive tract is the only one that is direct contact with food before is it digested?
   a. serosa
   b. longitudinal muscle
   * c. mucosa
   d. submucosa
   e. circular muscle

M **6.** Sphincters
   a. are circular muscles.
   b. prevent backflow.
   c. are smooth muscles.
   d. are found at the beginning and end of the stomach.
   * e. all of these

M **7.** Which process propels the food down the esophagus into the stomach?
   a. glycolysis
   b. plasmolysis
   c. emulsion
   * d. peristalsis
   e. all of these

E **8.** Which of the following layers forms the outer covering of the gastrointestinal tract?
   * a. serosa
   b. mucosa
   c. conjunctiva
   d. muscle layer
   e. submucosa

D **9.** The function of segmentation is to
   a. move the food through the digestive tract.
   b. churn the food and mix the contents with the digestive tract.
   c. bring the contents to the wall of the tract, where they could be absorbed.
   d. produce a wavelike push of the gut contents through the system.
   * e. both churn the food and mix the contents with the digestive tract plus bring the contents to the wall of the tract, where they could be absorbed.

CHEWING AND SWALLOWING

E **10.** Chewing
   a. breaks food down into smaller pieces.
   b. physically and mechanically breaks up the food.
   c. increases the surface area of food exposed to digestive enzymes.
   d. actually mixes some enzymes with the food.
   * e. all of these

E   11. The nerves and blood vessels of a human tooth are located in the
  a. dentine.
 * b. pulp.
  c. enamel.
  d. caries.
  e. periodontal membrane.

E   12. The digestion of which class of foods begins in the mouth?
 * a. carbohydrates
  b. proteins
  c. lipids
  d. amino acids
  e. nucleic acids

E   13 A bolus is formed in the
 * a. mouth.
  b. esophagus.
  c. stomach.
  d. small intestine.
  e. large intestine.

M   14. Salivary amylase is produced by all of the following EXCEPT the
  a. parotid glands.
  b. sublingual glands.
 * c. epiglottal glands.
  d. submandibular glands.

E   15. Which of the following organs of the digestive system is different from the other four because it does NOT produce any secretions that aid in the digestive process?
  a. stomach
  b. liver
 * c. esophagus
  d. pancreas
  e. salivary gland

M   16. During the process of swallowing, the
  a. esophagus is temporarily closed by the glottis.
 * b. epiglottis closes the trachea leading to the lungs.
  c. pharynx restricts food entry to the esophagus.
  d. epiglottis seals the esophagus.
  e. none of these

DIGESTION IN THE STOMACH AND SMALL INTESTINE

E   17. The digestion of proteins begins in the
 * a. stomach.
  b. pancreas.
  c. small intestine.
  d. large intestine.
  e. esophagus.

E   18. Chyme is formed in the
  a. mouth.
  b. esophagus.
 * c. stomach.
  d. small intestine.
  e. large intestine.

E   19. The acid released in the stomach is
  a. carbonic acid.
 * b. hydrochloric acid.
  c. nitric acid.
  d. sulfuric acid.
  e. phosphoric acid.

D   20. Which of the following is NOT an active form of an enzyme?
  a. trypsin
  b. amylase
  c. pepsin
 * d. pepsinogen
  e. chymotrypsin

M   21. Which of the following functions does the stomach perform the LEAST?
 * a. absorption
  b. digestion
  c. storage
  d. mixing
  e. movement

M   22. Which of the following is NOT a secretion of the stomach?
  a. pepsinogen
  b. mucus
  c. gastrin
  d. hydrochloric acid
 * e. lipase

M   23. High stomach acidity
  a. creates ideal conditions for carbohydrate digestion.
  b. promotes emulsification of fats.
 * c. favors protein digestion.
  d. blocks the release of histamine, thereby favoring production of peptic ulcers.
  e. converts lipases into their active forms.

M   24. Stomach motility
        a. decreases following a heavy meal.
      * b. controls the amount of material leaving
           the pyloric sphincter.
        c. is unaffected by emotional state or
           external environmental factors.
        d. may be retarded when stretch receptors
           on the stomach wall are activated.
        e. is increased by hormones released in
           response to high stomach acidity.

D   25. Digestion of the stomach wall by gastric
         juice is usually prevented by
        a. secretion of protein-digesting enzymes
           in inactive form.
        b. a covering of mucus.
        c. histamine.
      * d. secretion of protein-digesting enzymes
           in inactive form and a covering of
           mucus.
        e. secretion of protein-digesting enzymes
           in inactive form and a covering of
           mucus plus histamine

M   26. Which of the following factors does NOT
         stimulate the stomach to pass on its
         contents to the small intestine?
      * a. depression and fear
        b. stimulation of mechanoreceptors in the
           stomach wall following a large meal
        c. reduced fat or acid content of chyme in
           the duodenum
        d. elation and relaxation
        e. all of these

M   27. Which of the following speeds up the
         passage of food through the pyloric
         sphincter?
      * a. the larger number of mechanoreceptors
           in the stomach wall that are activated
        b. the presence of acid in the duodenum
        c. hormones such as cholecystokinin
        d. emotional conditions such as
           depression or fear
        e. the presence of fat in the duodenum

D   28. Which of the following components of a
         hamburger would leave the stomach
         chemically undigested?
        a. protein
        b. starch
        c. sugar (in the catsup)
        d. lipid
      * e. both sugar (in the catsup) and lipid

E   29. The first part of the small intestine is the
      * a. duodenum.
        b. ileum.
        c. colon.
        d. cecum.
        e. jejunum.

E   30. The digestion of fats mostly occurs in the
        a. stomach.
        b. pancreas.
      * c. small intestine.
        d. lymph vascular system.
        e. liver.

E   31. Of the following the greatest amount of
         nutrient absorption takes place in the
        a. stomach.
      * b. small intestine.
        c. colon.
        d. pancreas.
        e. esophagus.

D   32. A deficiency in the supply of pancreatic
         juice to the small intestine could have what
         result?
        a. lack of insulin
      * b. duodenal ulcers
        c. accumulation of large globs of fat
        d. lack of insulin and accumulation of
           large globs of fat
        e. lack of insulin, accumulation of large
           globs of fat, and duodenal ulcers

M   33. Concerning the role of the pancreas in
         digestion,
      * a. no digestion occurs in the pancreas.
        b. endocrine cells secrete bicarbonate,
           which helps neutralize highly acidic
           chyme.
        c. endocrine cells release enzymes that
           break down carbohydrates, fats,
           proteins, and nucleic acids in the small
           intestine.
        d. exocrine tissue produces insulin and
           glucagon that help regulate the
           metabolism of sugar.
        e. all of these

D   34. Which of the following performs a chemical
         digestion similar to that done by pepsin?
        a. aminopeptidase
        b. carboxypeptidase
      * c. trypsin
        d. gastrin
        e. lecithin

M   **35.** Which of the following does NOT digest
       proteins?
         a. trypsin
         b. chymotrypsin
         c. aminopeptidase
         d. pepsin
       * e. lipase

E   **36.** Ducts from the pancreas and liver enter the
         a. stomach.
         b. colon.
       * c. small intestine.
         d. gall bladder.
         e. rectum.

M   **37.** Which of the following is NOT found in
       bile?
         a. salts
         b. cholesterol
         c. pigments
       * d. digestive enzymes
         e. lecithin

M   **38.** Bile
         a. has no effect on digestion.
       * b. helps in the digestion of fats.
         c. helps in the digestion of carbohydrates.
         d. helps in the digestion of proteins.
         e. both helps in the digestion of
            carbohydrates and proteins.

E   **39.** Fats are digested by which of the following?
         a. aminopeptidase
         b. disaccharidases
         c. amylase
       * d. lipase
         e. trypsin

D   **40.** Which of the following are tiny projections
       of the mucosal wall?
         a. microvilli
         b. mucins
       * c. villi
         d. submucosa
         e. jejunum

ABSORPTION OF NUTRIENTS IN THE SMALL INTESTINE

M   **41.** Which of the following are absorbed by the
       lymphatic system?
         a. monosaccharides
         b. amino acids
         c. monoglycerides
         d. fatty acids
       * e. both monoglycerides and fatty acids

M   **42.** Movement of glucose through the
       membranes of the small intestine is
       primarily by
         a. osmosis.
         b. bulk flow.
       * c. active transport.
         d. diffusion.
         e. all of these

M   **43.** Micelles aid in the absorption of
         a. proteins.
         b. sugars.
         c. carbohydrates.
       * d. lipids.
         e. vitamins.

M   **44.** Which of the following is NOT absorbed
       directly into the blood from the intestine?
         a. water
       * b. triglycerides
         c. ions
         d. glucose
         e. amino acids

THE LIVER'S ROLE IN DIGESTION

M   **45.** The liver is assured "first choice" of all the
       nutrients absorbed by the intestine because
       of the
         a. hepatic vein.
         b. common bile duct.
         c. duodenum.
         d. hepatic artery.
       * e. hepatic portal vein.

E   **46.** The organ that stores and detoxifies different
       organic compounds is the
         a. pancreas.
         b. small intestine.
       * c. liver.
         d. spleen.
         e. gall bladder.

M   **47.** The liver is associated with all of the
       following functions EXCEPT
         a. formation of urea.
         b. formation of bile.
         c. detoxification of poisons.
       * d. secretion of bicarbonate ions.
         e. carbohydrate storage

M 48. The liver is associated with all of the following functions EXCEPT
   a. inactivation of drugs.
   b. assembly and storage of fats.
   c. assembly and disassembly of certain proteins.
   d. degradation of worn-out blood cells.
   * e. formation of glucagon.

M 49. Which organ takes glucose out of the blood and stores it as glycogen?
   a. pancreas
   b. spleen
   * c. liver
   d. skin
   e. kidney

THE LARGE INTESTINE

E 50. The primary function of the large intestine is
   a. storage of feces.
   b. retention of water.
   c. manufacture of vitamin K.
   d. digestion of fats.
   * e. absorption of water.

E 51. Which of these structures is located at the juncture of the small and large intestines?
   a. cardiac orifice
   * b. cecum
   c. anal canal
   d. lacteal
   e. rectum

D 52. Which of the following regions connects directly to the rectum?
   * a. sigmoid colon
   b. ascending colon
   c. descending colon
   d. transverse colon
   e. appendix

M 53. Bulk in the diet
   a. increases the length of time material is in the colon.
   b. increases the chance of cancer.
   * c. prevents diarrhea and irritable colon syndrome.
   d. may increase the incidence of appendicitis in the people who eat too much bulk.
   e. is characteristic of people in urban areas.

DIGESTION CONTROLS AND NUTRIENT TURNOVER

M 54. Which of the following acts enzymatically rather than hormonally?
   a. cholecystokinin
   * b. pepsin
   c. secretin
   d. gastrin
   e. glucose insulinotropic peptide

D 55. Which of the following is NOT a hormone?
   a. gastrin
   b. secretin
   * c. mucin
   d. cholecystokinin
   e. All of these are hormones.

M 56. Which of the following is NOT secreted by the intestinal mucosa?
   a. secretin
   b. cholecystokinin
   * c. gastrin
   d. GIP
   e. All are secreted by the intestine.

D 57. Which of the following chemicals is the first hormone secreted by the intestinal tract in response to the presence of food?
   a. salivary amylase
   b. cholecystokinin
   c. glucose insulinotropic peptide (GIP)
   * d. gastrin
   e. secretin

M 58. Which of the following secretes a hormone that causes its own secretory cells to respond?
   a. small intestine
   b. pancreas
   c. large intestine
   * d. stomach
   e. liver

THE BODY'S NUTRITIONAL REQUIREMENTS

M 59. The ideal diet consists of all of the following EXCEPT
   a. bulk.
   * b. few complex carbohydrates.
   c. little salt and sugar.
   d. little red meat.
   e. fish, poultry, and legumes.

M 60. The surgeon general recommends reducing all but which of the following components of our diet?
   a. saturated fat
   b. cholesterol
 * c. complex carbohydrates
   d. salt
   e. sugar

M 61. Which of the following should be present in the human diet in the highest percentage?
   a. protein
 * b. carbohydrate
   c. lipid
   d. vitamins
   e. minerals

M 62. During, or shortly after a meal, most cells use which of the following as a source of energy?
   a. fat
   b. amino acids
 * c. glucose
   d. glycogen
   e. any of the above, depending on the concentration of the particular organic compound

E 63. Of the 20 amino acids, how many are considered to be essential in that the human body cannot synthesize them?
   a. 2
   b. 5
 * c. 8
   d. 10
   e. 12

M 64. Which of the following represents the best source of protein (net protein utilization)?
   a. milk
 * b. eggs
   c. fish
   d. soybeans
   e. cheese

D 65. People who do not eat meat (vegetarians) must choose their food carefully to get the necessary
   a. vitamins.
   b. minerals.
   c. carbohydrates.
 * d. amino acids.
   e. fatty acids.

M 66. Lipids can serve in all but which of the following capacities?
 * a. enzymes
   b. energy
   c. membrane structure
   d. insulation
   e. It can serve in all of these capacities.

VITAMINS AND MINERALS

M 67. Which of the following vitamins is fat-soluble and can be stored in the body?
 * a. A
   b. B_1 (thiamine)
   c. C (ascorbic acid)
   d. B_2 (riboflavin)
   e. niacin

E 68. Which vitamin functions in forming a blood clot?
   a. A
   b. E
 * c. K
   d. B
   e. all of these

M 69. Which of the following statements is NOT true concerning mineral metabolism?
   a. Sodium and potassium are needed for maintaining osmotic balances.
 * b. Zinc is important in building strong bones and teeth.
   c. Sodium and potassium are needed for muscle and nerve functioning.
   d. Iron is needed for building cytochromes and heme groups.
   e. all the statements are true

E 70. Lack of which element can lead to thyroid problems?
   a. iron
 * b. iodine
   c. calcium
   d. zinc
   e. magnesium

M 71. The element needed for blood clotting, nerve transmission, and bone and tooth formation is
   a. iron.
   b. iodine.
 * c. calcium.
   d. zinc.
   e. magnesium.

M   72. A deficiency of which vitamin produces rickets in children and osteomalcia in adults?
     a. A
     b. B
     c. C
   * d. D
     e. E

M   73. A deficiency of vitamin C may give rise to
     a. beriberi.
   * b. scurvy.
     c. pellagra.
     d. hypothyroidism.
     e. all of these

M   74. Beriberi is a deficiency disease related to which of the following vitamins?
     a. A
   * b. B_1 (thiamine)
     c. C (ascorbic acid)
     d. B_2 (riboflavin)
     e. niacin

M   75. Pellagra is a deficiency disease related to which of the following vitamins?
     a. A
     b. B_1 (thiamine)
     c. C (ascorbic acid)
     d. B_2 (riboflavin)
   * e. niacin

M   76. Scurvy is a deficiency disease related to which of the following vitamins?
     a. A
     b. B_1 (thiamine)
   * c. C (ascorbic acid)
     d. B_2 (riboflavin)
     e. niacin

M   77. Lack of which element can lead to goiter?
     a. iron
   * b. iodine
     c. calcium
     d. zinc
     e. magnesium

E   78. The constituent of hemoglobin whose absence leads to anemia is
   * a. iron.
     b. iodine.
     c. calcium.
     d. zinc.
     e. magnesium.

FOOD ENERGY AND BODY WEIGHT

E   79. Body weight is controlled by
     a. caloric intake.
     b. energy utilization.
     c. level of metabolism.
     d. age and sex.
   * e. all of these

E   80. Obesity is defined as what percent over the ideal weight?
     a. 10
     b. 15
     c. 20
     d. 25
   * e. 30

E   81. Obese people have a greater risk of
     a. diabetes.
     b. atherosclerosis.
     c. high blood pressure.
     d. death at all ages.
   * e. all of these

E   82. If caloric intake is balanced with energy output, body
     a. weight gain will occur.
     b. weight loss will occur.
   * c. weight will remain stable.
     d. fat content will increase.
     e. protein will decrease.

CHOICES: BIOLOGY AND SOCIETY: MALNUTRITION AND UNDERNUTRITION

     83. Which of the following does NOT involve a nutritional deficiency?
     a. anoxia
   * b. obesity
     c. xeropthalmia
     d. kwashiorkor
     e. scurvy

Matching Questions

D   **84.** Choose the one most appropriate answer for each.

1 _____ amylase

2 _____ bile

3 _____ disaccharidase

4 _____ gastrin

5 _____ lipase

6 _____ monosaccharides

7 _____ pepsin

8 _____ amino- and carboxypeptidase

9 _____ sphincter

10 _____ trypsin

    A. made in the small intestine and pancreas; acts on protein fragments

    B. glucose, fructose, and galactose

    C. made in the pancreas; acts on fats

    D. made by the pancreas and salivary glands; acts on starch

    E. made by the small intestine; acts on double sugars

    F. made by the pancreas; acts on proteins and polypeptides

    G. contains cholesterol; helps emulsify fats

    H. made in the stomach; acts on proteins

    I. stimulates hydrochloric acid secretion

    J. separates the stomach from the small intestine

Answers:

1. D	2. G	3. E	
4. I	5. C	6. B	
7. H	8. A	9. J	
10. F			

Classification Questions

Answer questions 85–89 in reference to the five components of the gastrointestinal tract listed below:

    a. stomach
    b. gall bladder
    c. small intestine
    d. appendix
    e. large intestine

D   **85.** Many organisms, such as birds, have ceca (digestive pouches) in which bacteria break down difficult to digest plant materials. These ceca are homologous to which of the above in humans?

M   **86.** This organ absorbs about 95 percent of the water that enters the human body, either as fluids or as part of food being eaten.

M   **87.** Enzymatic digestion of proteins occurs primarily in this organ.

E   **88.** Bile salts, bile pigments, cholesterol, and lecithin are stored by this organ.

E   **89.** The digestion of proteins begins in this part of the digestive system.

Answers:

85. d	86. e	87. c
88. b	89. a	

Answer questions 90–94 in reference to the four glands or structures of the mammalian gastrointestinal tract listed below:

    a. salivary glands
    b. stomach lining
    c. intestinal lining
    d. pancreas

M   **90.** This is where the enzyme pepsin is produced.

D   **91.** This is where the enzyme carboxypeptidase is produced.

D   **92.** This is where the fat digesting enzyme, lipase, is formed.

D   **93.** This is where the peptide digesting enzyme, aminopeptidase, is produced.

M   **94.** This is where the protein digesting enzyme, trypsin, is produced.

Answers:

90. b	91. d	92. d
93. c	94. d	

Answer questions 95–99 in reference to the five vitamins listed below:

    a. Vitamin B_1
    b. Vitamin B_2
    c. Niacin
    d. Vitamin B_6
    e. Vitamin B_{12}

D   **95.** This vitamin is a coenzyme involved in amino acid metabolism and obtained from meat, potatoes, even spinach.

D   **96.** This vitamin is a component of the coenzyme thiamine pyrophosphate.

M   **97.** This vitamin is a constituent of the coenzymes NAD^+ and $NADP^+$.

M   **98.** This vitamin is more commonly known as riboflavin.

D   **99.** This vitamin acts as a coenzyme in nucleic acid metabolism.

Answers: 95. d   96. a   97. c
               98. b   99. e

Selecting the Exception

E   **100.** Four of the five answers listed below are structures through which ingested foodstuffs travel. Select the exception.
    a. crop
 * b. liver
    c. gizzard
    d. rectum
    e. small intestine

M   **101.** Four of the five answers listed below are functions of the digestive organs. Select the exception.
 * a. excretion
    b. absorption
    c. motility
    d. secretion
    e. digestion

M   **102.** Four of the five answers listed below release secretions that assist digestion. Select the exception.
    a. salivary gland
 * b. esophagus
    c. pancreas
    d. gall bladder
    e. liver

M   **103.** Four of the five answers listed below increase HCl secretion. Select the exception.
    a. cola drinks
    b. coffee
 * c. beer
    d. chocolate
    e. tea

D   **104.** Four of the five answers listed below are end products of digestion ready for intestinal absorption. Select the exception.
    a. monoglyceride
    b. nucleotides
 * c. disaccharides
    d. amino acids
    e. free fatty acids

M   **105.** Four of the five answers listed below digest the same class of foods. Select the exception.
    a. aminopeptidase
    b. pepsin
 * c. amylase
    d. trypsin
    e. chymotrypsin

M   **106.** Four of the five answers listed below perform their task in the same workplace. Select the exception.
 * a. pepsin
    b. lipase
    c. carboxypeptidase
    d. nuclease
    e. disaccharidase

M   **107.** Four of the five answers listed below are conditions related to diet. Select the exception.
    a. colon cancer
    b. kidney stones
    c. cardiovascular disorders
    d. obesity
 * e. Alzheimer's disease

M   **108.** Four of the five answers listed below are all of the same class of vitamins. Select the exception.
    a. Vitamin A
    b. Vitamin K
 * c. Vitamin C
    d. Vitamin D
    e. Vitamin E

D 109. Four of the five answers listed below are functions performed by the same organ. Select the exception.
* a. regulates pH of body fluids
  b. removes toxic substances from the blood
  c. storage and interconversion of carbohydrates, fats, and proteins
  d. inactivates hormones
  e. formation of urea from nitrogenous waste

M 110. Four of the five answers listed below are layers of the digestive tract. Select the exception.
* a. peritoneum
  b. mucosa
  c. serosa
  d. submucosa
  e. muscle layer

D 111. Four of the five answers listed below are hormones associated with digestion. Select the exception.
* a. bile
  b. gastrin
  c. secretin
  d. cholecystokinin
  e. glucose insulinotropic peptide

D 112. Four of the five answers listed below are conditions caused by vitamin deficiency. Select the exception.
  a. scurvy
  b. pellagra
  c. rickets
  d. beriberi
* e. goiter

M 113. Four of the five answers listed below are members of the B complex vitamins. Select the exception.
* a. ascorbic acid
  b. folic acid
  c. pantothenic acid
  d. riboflavin
  e. thiamine

M 114. Four of the five answers below are associated with digestive disorders. Select the exception.
  a. Crohn's disease
* b. basal metabolism
  c. lactose intolerance
  d. cystic fibrosis
  e. food allergies

CHAPTER 7
BLOOD AND CIRCULATION

Multiple-Choice Questions

BLOOD

M   **1.** Which cell is NOT the same type as the others?
  * a. erythrocytes
    b. neutrophils
    c. lymphocytes
    d. eosinophils
    e. monocytes

M   **2.** Which cell is NOT involved with the defense response?
  * a. erythrocytes
    b. neutrophils
    c. lymphocytes
    d. eosinophils
    e. monocytes

M   **3.** Which cell is the most abundant in the human body?
    a. lymphocytes
    b. basophils
  * c. erythrocytes
    d. neutrophils
    e. platelets

M   **4.** Which cell produces the fibrin used in blood clots?
    a. lymphocytes
    b. basophils
    c. erythrocytes
    d. neutrophils
  * e. platelets

E   **5.** About how many quarts of blood does a normal, 150 pound, human adult have?
    a. 1-2
    b. 3-4
  * c. 4-5
    d. 5-6
    e. 6-7

E   **6.** What percent of the total blood volume is plasma?
    a. 15 to 25
    b. 33 to 40
  * c. 50 to 60
    d. 66 to 75
    e. about 80

M   **7.** Megakaryocytes fragment to produce
    a. red blood cells.
    b. lymphocytes.
  * c. platelets.
    d. eosinophils.
    e. neutrophils.

E   **8.** All but which of the following can occur in the blood?
  * a. digestion of nutrients
    b. combining of oxygen with hemoglobin
    c. transport of phagocytic cells
    d. stabilization of pH
    e. equalization of internal temperatures

E   **9.** The most common plasma protein is
  * a. albumin.
    b. fibrin.
    c. fibrinogen.
    d. gamma globulin.
    e. hemoglobin.

M   **10.** The plasma protein associated with immunity is
    a. alpha globulin.
    b. beta globulin.
  * c. gamma globulin.
    d. albumin.
    e. fibrinogen.

E   **11.** If a test tube of whole blood is subjected to centrifugation, the cells will be packed in the bottom of the tube and the fluid above it will be designated
    a. water.
    b. serum.
    c. lymph.
    d. interstitial fluid.
  * e. plasma.

OXYGEN TRANSPORT IN BLOOD

E   **12.** Most of the oxygen in the blood is transported by
    a. plasma.
    b. serum.
    c. platelets.
  * d. hemoglobin.
    e. leukocytes.

E　13. Hemoglobin contains which element?
　　　　　a. chlorine
　　　　　b. sodium
　　*　c. iron
　　　　　d. copper
　　　　　e. magnesium

E　14. Blood rich in oxygen is what color?
　　　　　a. yellow
　　　　　b. pink
　　*　c. bright red
　　　　　d. blue
　　　　　e. purple

D　15. All of the following promote the unloading of oxygen at needy tissues EXCEPT
　　　　　a. high metabolism.
　　*　b. abundant oxygen supply.
　　　　　c. higher temperatures.
　　　　　d. lower pH.
　　　　　e. small diameter blood vessels.

LIFE CYCLE OF RED BLOOD CELLS

E　16. In adult humans, red blood cells originate in the
　　　　　a. liver.
　　　　　b. spleen.
　　　　　c. kidneys.
　　*　d. bone marrow.
　　　　　e. thymus gland.

E　17. In humans, which cell does NOT have a nucleus when mature?
　　*　a. erythrocytes
　　　　　b. lymphocytes
　　　　　c. neutrophils
　　　　　d. eosinophils
　　　　　e. monocytes

M　18. How long does the average red blood cell live?
　　　　　a. 4 days
　　　　　b. 4 weeks
　　*　c. 4 months
　　　　　d. 1 year
　　　　　e. 4 years

M　19. Stem cells
　　*　a. retain the ability to divide and give rise to groups of cells.
　　　　　b. are phagocytic.
　　　　　c. are the most common blood cells.
　　　　　d. transport oxygen and carbon dioxide.
　　　　　e. are more numerous in women than in men.

D　20. Erythropoietin
　　　　　a. is the most common blood protein.
　　　　　b. is secreted by the kidney.
　　*　c. stimulates the red bone marrow to generate red blood cells.
　　　　　d. is an enzyme that functions in blood clotting.
　　　　　e. increases the ability of plasma to transport carbon dioxide.

D　21. When the oxygen level in the tissues is low, which of the following secretes enzymes that trigger the production of erythropoietin, causing an increase in red blood cell production?
　　　　　a. lungs
　　*　b. kidneys
　　　　　c. liver
　　　　　d. spleen
　　　　　e. pancreas

M　22. When red blood cells are destroyed, which one of the following is NOT recycled by the body?
　　　　　a. iron
　　　　　b. amino acids
　　*　c. heme
　　　　　d. iron and amino acids.
　　　　　e. iron and heme

E　23. Which of the following is NOT a hereditary anemia?
　　*　a. iron-deficiency anemia
　　　　　b. pernicious anemia
　　　　　c. sickle-cell anemia
　　　　　d. thalassemia

BLOOD TYPING

D　24. If you are blood type A,
　　*　a. you carry antibodies for type B blood.
　　　　　b. you carry markers for type B blood.
　　　　　c. you can donate blood to a person with type O blood.
　　　　　d. you can receive blood from a person with type AB blood.
　　　　　e. none of these

M　25. Type A blood will NOT agglutinate when mixed with
　　　　　a. type B blood.
　　　　　b. type A blood.
　　　　　c. type AB blood.
　　　　　d. type O blood.
　　*　e. both A and AB, but will clump with types B and O.

E   **26.** Which blood type is the universal donor?
- a. A+
- b. B
- c. AB+
- d. AB
- *   e. O

E   **27.** Which blood type is the universal recipient?
- a. A
- b. B+
- c. AB+
- *   d. AB
- e. O+

D   **28.** In the Rh disease
- a. the mother must be positive and her first and second children positive.
- *   b. the mother must be negative and her first and second children positive.
- c. the mother must be negative and her first and second children negative.
- d. the mother must be positive and her first and second children negative.
- e. the mother and the father must both be negative and the child positive.

OVERVIEW OF THE CARDIOVASCULAR SYSTEM

D   **29.** Which of the following systems is the only one to have direct interactions with the other three?
- a. digestive
- b. urinary
- *   c. circulatory
- d. respiratory

M   **30.** Blood in arteries
- *   a. always travels away from the heart.
- b. travels away from the heart only if it is oxygen-rich.
- c. always travels toward the heart.
- d. travels from the lungs.
- e. is always oxygen-rich.

E   **31.** Blood moves most slowly through
- a. arteries.
- *   b. capillaries.
- c. venules.
- d. veins.
- e. arterioles.

THE HEART: A DURABLE PUMP

M   **32.** Atria differ from ventricles in that they
- a. are larger.
- b. have thicker walls with more muscles.
- *   c. receive blood from veins.
- d. have a higher blood pressure.
- e. empty through the semilunar valves.

M   **33.** The coronary vessels
- *   a. supply and drain the heart muscle.
- b. bypass the heart ventricles.
- c. send blood directly to the lungs.
- d. are not really necessary because the heart can get its blood supply from the "inside."
- e. lead directly from the atria to the ventricles.

E   **34.** The receiving zone of a vertebrate heart is
- a. a plaque.
- b. the aorta.
- *   c. an atrium.
- d. a capillary bed.
- e. all of these

D   **35.** The bicuspid (mitral) valve is located between the
- a. left and right atria.
- b. right atrium and right ventricle.
- c. the left and right ventricles.
- *   d. left atrium and left ventricle.
- e. left ventricle and the aorta.

M   **36.** The cardiac muscle of the heart is called the
- a. pericardium.
- b. endocardium.
- c. endothelium.
- d. chordae tendineae.
- *   e. myocardium.

M   **37.** What occurs during systole?
- a. Oxygen-rich blood is pumped to the lungs.
- *   b. The heart muscle tissues contract.
- c. The atrioventricular valves suddenly open.
- d. Oxygen-poor blood from all body regions except the lungs flows into the right atrium.
- e. all of these

D   **38.** If a physician hears two "lub" sounds instead of one, then which of the following conditions is true?
- a. The semilunar valves are not closing simultaneously.
- b. The atrial blood is flowing backward and causing the extra sound.
- * c. The atrioventricular valves are not closing at the same time.
- d. The AV and semilunar valves are not closing at the same time.
- e. No such double sound has ever been heard.

BLOOD CIRCULATION

M   **39.** The pulmonary circulation
- a. involves the hepatic portal vein.
- b. moves oxygen-rich blood to the kidneys.
- c. includes the coronary arteries.
- * d. leads to, through, and from the lungs.
- e. all of these

M   **40.** In the human systemic circuit, blood will pass through all but which of the following?
- a. liver
- b. limbs
- * c. lungs
- d. digestive organs
- e. brain

D   **41.** Which of the following statements is true?
- a. Arteries carry only oxygenated blood.
- * b. The systemic circuit leaves the heart from the left ventricle.
- c. Blood passes through only one capillary bed on its trip through the systemic circuit.
- d. Platelets survive a longer time than erythrocytes.
- e. The heart is able to pick up the oxygen it needs as the blood flows through it.

M   **42.** The aorta leaves the
- a. left atrium.
- b. right atrium.
- * c. left ventricle.
- d. right ventricle.

M   **43.** The pulmonary artery carries blood away from the
- a. aorta.
- b. right atrium.
- * c. right ventricle.
- d. left atrium.
- e. left ventricle.

E   **44.** Blood from the body is first received by the heart in the
- a. coronary vein.
- b. left atrium.
- c. right ventricle.
- * d. right atrium.
- e. left ventricle.

HOW THE HEART CONTRACTS

M   **45.** Heart excitation originates in the
- a. atrioventricular node.
- b. intercalated disk.
- * c. sinoatrial node.
- d. pericardium.
- e. all of these

D   **46.** An artificial pacemaker supplements the actions of
- a. sympathetic nerves.
- b. the atrioventricular node.
- c. the medulla oblongata.
- * d. the sinoatrial node.
- e. the heart muscle itself.

M   **47.** Which of the following statements is false?
- * a. A heart will stop beating when the nerves to the heart are severed.
- b. Some cardiac muscle cells are self-excitatory.
- c. The pacemaker of the heart is the sinoatrial node.
- d. Cardiac muscles join end to end to allow rapid communication.
- e. Cardiac muscles contract essentially in unison.

M   48. The heart
     *   a. will contract as a result of stimuli from
              the sinoatrial node.
          b. contracts only as a result of nerve
              stimulation from the central nervous
              system.
          c. is activated primarily through the
              autonomic nervous system.
          d. pulse is primarily under the control of
              the atrioventricular node.
          e. is completely independent of all
              nervous control.

BLOOD PRESSURE AND VELOCITY IN THE CARDIOVASCULAR SYSTEM

E   49. Which of the following has the highest
          blood pressure?
          a. right ventricle
          b. right atrium
     *   c. left ventricle
          d. left atrium
          e. pulmonary circulation

M   50. The diastolic pressure for a normal young
          adult would be
          a. 60 mmHg.
     *   b. 80 mmHg.
          c. 100 mmHg.
          d. 120 mmHg.
          e. 140 mmHg.

E   51. The part of the brain responsible for blood
          pressure is the
     *   a. medulla oblongata.
          b. cerebellum.
          c. cerebrum.
          d. corpus callosum.
          e. pons.

M   52. Blood pressure is highest in the
     *   a. aorta.
          b. pulmonary artery.
          c. capillary bed.
          d. subclavian vein.
          e. lower vena cava.

HOW VESSEL STRUCTURE AFFECTS THE PRESSURE AND FLOW OF BLOOD

M   53. Which of the following is NOT found in an
          arteriole?
          a. elastic layer
          b. basement membrane
          c. smooth muscle
     *   d. valve
          e. endothelium

E   54. By controlling their musculature, which
          vessels can vary the resistance to blood
          flow?
          a. arteries
          b. veins
          c. capillaries
     *   d. arterioles
          e. all of these

E   55. In its travel through the human body, blood
          usually continues on from capillaries to
          enter
          a. arterioles.
     *   b. venules.
          c. arteries.
          d. veins.
          e. other capillaries.

M   56. The greatest drop in blood pressure occurs
          in the
          a. arteries.
     *   b. arterioles.
          c. capillaries.
          d. venules.
          e. veins.

E   57. By controlling their musculature, which of
          the following can vary the resistance to
          blood flow?
          a. arteries
          b. veins
          c. capillaries
     *   d. arterioles
          e. all of these

M   58. Because of their great elasticity, which of
          the following can function as blood volume
          reservoirs during times of low metabolic
          output?
     *   a. veins and venules
          b. arteries
          c. arterioles
          d. capillaries
          e. all of these

E   **59.** Which controls the distribution of blood?
     a. arteries
   *   b. arterioles
     c. capillaries
     d. venules
     e. veins

E   **60.** Which are pressure reservoirs with low resistance to flow?
   *   a. arteries
     b. arterioles
     c. capillaries
     d. venules
     e. veins

E   **61.** Which are highly distensible reservoirs for blood volume?
     a. arteries
     b. arterioles
     c. capillaries
     d. venules
   *   e. veins

E   **62.** The greatest volume of blood is found in the
     a. aorta and arteries.
     b. capillaries.
   *   c. veins.
     d. lungs.
     e. heart.

A CLOSER LOOK AT CAPILLARIES

M   **63.** The blood makes "pickups" and "deliveries" MOST DIRECTLY to
     a. body cells.
     b. lymph.
   *   c. interstitial fluid.
     d. non-blood tissue.
     e. organs.

M   **64.** The interstitial fluid is
     a. a reservoir.
     b. the extracellular fluid.
     c. supplied by the blood.
     d. similar to sea water.
   *   e. all of these

M   **65.** Extracellular fluid contains all but which of the following?
   *   a. erythrocytes
     b. ions
     c. white blood cells
     d. lymph
     e. water

M   **66.** At the arteriole end of the capillary, more fluid leaves the capillary than enters as a result of
     a. osmotic pressure.
   *   b. hydrostatic force.
     c. gap junctions.
     d. vasodilation.
     e. all of these

HEMOSTASIS AND BLOOD CLOTTING

D   **67.** Hemostasis in vertebrates includes all of the following EXCEPT
     a. blood clot formation.
     b. vessel constriction.
   *   c. release of iron to aid in the clumping of platelets.
     d. vessel spasms.
     e. platelets releasing substances that cause them to attract each other.

M   **68.** Which of the following is NOT involved in the formation of a blood clot?
   *   a. plasma cells
     b. fibrinogen
     c. thrombin
     d. fibrin
     e. All of these are involved.

M   **69.** Which of the following is an enzyme?
     a. fibrinogen
     b. fibrin
     c. prothrombin
   *   d. thrombin
     e. plasminogen

FOCUS ON YOUR HEALTH: CARDIOVASCULAR DISORDERS

M   **70.** The most common vascular disease is
     a. phlebitis.
   *   b. hypertension.
     c. leukemia.
     d. sickle cell anemia.
     e. a stroke.

E   **71.** A stroke is a rupture of a blood vessel in the
     a. leg.
   *   b. brain.
     c. heart.
     d. lung.
     e. internal organs.

M   72. In atherosclerosis
     a. abnormal multiplication of smooth
        muscle cells in blood vessels occurs.
     b. the arterial walls fill with connective
        tissue.
     c. the lipids in the bloodstream become
        embedded in the walls of the endothelial
        lining.
     d. a fibrous net covers the entire abnormal
        area.
  * e. all of these

M   73. The mineral associated with atherosclerosis
         is
     a. iron.
     b. magnesium.
     c. cobalt.
  * d. calcium.
     e. iodine.

M   74. Cholesterol is believed to be carried by
     a. albumin.
     b. high-density lipoproteins.
     c. low-density lipoproteins.
     d. triglycerides.
  * e. both high-density and low-density
        lipoproteins.

THE LYMPHATIC SYSTEM

E   75. Which of the following is NOT a function
         of the lymph system?
     a. fighting infection
  * b. transporting dissolved gases
     c. reclaiming fluids
     d. harboring white blood cells
     e. All are functions of the lymph system.

M   76. Which of the following is transported in
         greater quantities in the lymphatic system
         than in the blood?
     a. red blood cells
     b. wastes
  * c. fats
     d. amino acids
     e. white blood cells

M   77. Which statement is NOT true of the lymph
         vascular system? The lymph vascular
         system
     a. transports lipids absorbed from the
        small intestine to the bloodstream.
     b. recovers and transports interstitial fluid
        back to the bloodstream.
  * c. absorbs glucose from the small
        intestine and transports it to the brain.
     d. serves the body's system of defenses
        against bacteria and other infectious
        agents.
     e. performs all of these functions.

M   78. The lymphoid organs include all but the
     a. spleen.
  * b. stomach.
     c. thymus.
     d. tonsils
     e. appendix.

E   79. The system which reclaims fluids and
         proteins that have escaped from blood
         capillaries is the
     a. cardiovascular.
     b. pulmonary.
  * c. lymphatic.
     d. sinoatrial.
     e. venous.

E   80. Areas where lymphocytes congregate as
         they cleanse the blood of foreign materials
         are called
     a. stem cells.
     b. SA nodes.
     c. capillary beds.
  * d. lymph nodes.
     e. antibodies.

E   81. Which lymphoid organ serves as a reservoir
         for lymphocytes?
  * a. spleen
     b. thymus
     c. tonsils
     d. lymph nodes
     e. appendix

Classification Questions

Answer questions 82–86 in reference to the five components of mammalian blood listed below:

    a. red blood cells
    b. basophils
    c. platelets
    d. serum albumin protein
    e. sodium and potassium chloride

E   **82.** This blood component plays a central role in clotting blood following a wound.

E   **83.** This blood component contains hemoglobin.

D   **84.** This blood component plays a role in the inflammatory response and allergic responses.

E   **85.** This blood component plays a role in maintaining the ionic balance of the body.

E   **86.** Oxygen is transported throughout the body by this blood component.

Answers: 82. c   83. a   84. b
              85. e   86. a

Answer questions 87–91 in reference to the four structures of the heart listed below:

    a. right atrium
    b. left atrium
    c. left ventricle
    d. right ventricle

M   **87.** Blood from the upper and lower vena cavas enters the heart via this structure.

M   **88.** Blood passes to the lungs from this structure.

M   **89.** Deoxygenated blood exits the heart from this structure.

M   **90.** Oxygenated blood enters the heart via this structure.

M   **91.** Blood is pumped to the majority of the body by this structure.

Answers: 87. a   88. d   89. d
              90. b   91. c

Selecting the Exception

D   **92.** Four of the five answers listed below are related by the "type" of blood they carry. Select the exception.
        a. pulmonary artery
   *   b. aorta
        c. hepatic portal vein
        d. inferior vena cava
        e. jugular vein

M   **93.** Four of the five answers listed below are related by a common property. Select the exception.
        a. neutrophil
   *   b. erythrocyte
        c. lymphocytes
        d. monocyte
        e. basophil

M   **94.** Four of the five answers listed below are blood proteins. Select the exception.
   *   a. epinephrine
        b. globulin
        c. hemoglobin
        d. fibrinogen
        e. albumin

M   **95.** Four of the five answers listed below are characteristics of most veins. Select the exception.
        a. blood volume reservoir
        b. contain valves
        c. low resistance-transport tubes
   *   d. transport oxygen
        e. low blood pressure

M   **96.** Three of the four answers listed below are related by a common function. Select the exception.
   *   a. gamma globulin
        b. prothrombin
        c. fibrin
        d. fibrinogen

D   **97.** Four of the five answers listed below are related by a common feature. Select the exception.
        a. A
        b. B
        c. AB
   *   d. Rh+
        e. O

D   **98.** Four of the five answers listed below are related by a common function. Select the exception.

   *   a. heart
       b. spleen
       c. thymus
       d. tonsils
       e. lymph node

CHAPTER 8
IMMUNITY

Multiple-Choice Questions

DESPERATE MEASURES FOR DESPERATE TIMES

E   1. The first disease for which a successful vaccination was developed was
    a. the plague.
*   b. smallpox.
    c. rabies.
    d. chicken pox.
    e. diphtheria.

E   2. The person who developed and demonstrated the first successful vaccine was
    a. Pasteur.
    b. Koch.
    c. Lister.
*   d. Jenner.
    e. Erhlich.

E   3. The word *vaccination* comes from the Latin word for
    a. germ.
*   b. cow.
    c. chicken.
    d. rabbit.
    e. rat.

THREE LINES OF DEFENSE

M   4. All but which of the following can be called a pathogen?
    a. virus
    b. bacterium
    c. fungus
*   d. cancer
    e. protozoan

M   5. The barrier to invasion by microbes involves
    a. stomach acids.
    b. the symbiotic microorganisms already in the body.
    c. ciliated mucous membranes.
    d. lysozyme and other enzymes.
*   e. all of these

E   6. All but which of the following are good barriers to invasion by microbes?
    a. mucous membranes
    b. eye secretions
*   c. broken skin
    d. stomach juices
    e. gut bacteria

M   7. Lysozyme
    a. is secreted by endocrine glands in the skin.
*   b. destroys the cell wall of invading bacteria.
    c. is produced in the lymph nodes and actively disables bacteria.
    d. has proved to be a very effective defense against viruses.
    e. is active within the circulatory system.

D   8. Normal bacterial inhabitants of the human body
    a. are naturally resistant to antibiotics.
*   b. are able to outcompete some invading pathogens and thus are one of the body's defense mechanisms.
    c. can be transformed into pathogenic forms if a person's resistance to disease is low.
    d. are unable to survive the human body's defense mechanisms.
    e. none of these

COMPLEMENT PROTEINS

M   9. Which system involves plasma proteins activated when they contact a bacterial cell?
    a. infection
*   b. complement
    c. bodyguard
    d. enhancer
    e. defender

D   10. Which of the following would NOT be an action of the complement system?
    a. lysis of a pathogen's membrane
*   b. trapping of pathogens in tangled protein threads
    c. marking of pathogens for destruction by macrophages
    d. attraction of phagocytes to scene of pathogen invasion

M 11. The complement system
    a. includes a group of about 20 plasma proteins.
    b. induces a cascade of proteins that counteract invasion by coating the invading cells.
    c. attracts phagocytic leukocytes to attack invading cells.
    d. causes the lysis of the plasma membranes of invading cells.
*   e. all of these

INFLAMMATION

M 12. White blood cells are derived from stem cells in the
    a. spleen.
    b. thymus.
*   c. bone marrow.
    d. blood.
    e. liver.

M 13. All of the following are white blood cells EXCEPT
    a. monocytes.
*   b. erythrocytes.
    c. neutrophils.
    d. eosinophils.
    e. macrophages.

M 14. Phagocytes perform their services in
    a. the blood.
    b. tissue spaces.
    c. the lymph system.
    d. the blood and tissue spaces only.
*   e. the blood, tissue spaces, and lymph system.

E 15. The accumulation of fluid at the site of a wound is the result of the secretion of
    a. kinins.
*   b. histamines.
    c. neutrophils.
    d. interferons.
    e. leukocytes.

M 16. Histamine causes
    a. blood vessels to contract.
    b. capillaries to lose their permeability.
*   c. an outward flow of fluids from the capillaries.
    d. a destruction of mast cells.
    e. an opening of the area of infection, through which the body's defense system can enter.

M 17. Inflammation
*   a. leads to the release of histamine, which causes capillaries to become "leaky."
    b. is accentuated by the administration of antihistamine drugs.
    c. does not occur during allergic reactions.
    d. is initiated by the buildup of dead cells and bacteria.
    e. is not affected by the action of the complement system.

M 18. Which event does NOT occur in the inflammatory response?
    a. Tissue swells because of outflow from capillary beds.
*   b. Blocking antibodies inactivate the resident mast cells.
    c. White blood cells are attracted to the area by chemotaxis.
    d. Complement proteins help identify invading material.
    e. The foreign invaders are engulfed and destroyed by phagocytosis.

D 19. An antihistamine drug would have as its SPECIFIC action which of the following?
*   a. constriction of the capillaries
    b. decreased redness
    c. reduced warmth
    d. promotion of clotting
    e. attraction of phagocytes

M 20. Which of the following is a secretion that regulates the interactions of white blood cells?
    a. antigen
    b. antibody
*   c. interleukin
    d. edema
    e. histamine

D 21. Interleukins
    a. are secreted by macrophages.
    b. trigger any B cell that has become sensitive to the specific antigen (the one inducing interleukin production) to divide.
    c. are the chemical triggers that cause tissue to release antihistamine.
    d. are effective only on pathogens that have invaded body cells.
*   e. are secreted by macrophages and trigger any B cell that has become sensitive to the specific antigen (the one inducing interleukin production) to divide.

THE IMMUNE SYSTEM

M   **22.** Terms that describe the immune response include all of the following EXCEPT
- a. specific.
- b. rapid.
- c. memory.
- * d. general.
- e. effective.

E   **23.** Which cells are divided into two groups: T cells and B cells?
- a. macrophages
- * b. lymphocytes
- c. complement cells
- d. platelets
- e. all of these

M   **24.** Which cells produce and secrete antibodies that set up bacterial invaders for subsequent destruction by macrophages?
- a. phagocytes
- b. macrophages
- * c. B cells
- d. T cells
- e. all of these

M   **25.** Which of the following are NOT generally targets of T cells?
- a. transplants of foreign tissue
- b. cancer
- c. infections caused by viruses
- * d. infections caused by bacteria
- e. all of these

M   **26.** Which cells produce antibodies?
- a. helper T
- b. suppressor T
- c. cytotoxic T
- d. natural killer
- * e. B

M   **27.** Which cells cause rapid division of the lymphocytes?
- * a. helper T
- b. suppressor T
- c. cytotoxic T
- d. memory
- e. B

M   **28.** Which cells are held in reserve to be used for a rapid response to subsequent intruders of the same type?
- a. helper T
- b. suppressor T
- c. cytotoxic T
- * d. memory
- e. B

M   **29.** Which cells directly destroy body cells infected by viral or fungal parasites?
- a. helper T
- b. suppressor T
- * c. cytotoxic T
- d. memory
- e. B

M   **30.** Mutant and cancerous cells are destroyed by which cells?
- a. helper T
- b. suppressor T
- * c. cytotoxic T
- d. memory
- e. B

D   **31.** Which of the following statements is false?
- a. Cytotoxic T cells kill cancer cells only if the cause is viral.
- b. MHC markers of grafted cells are identified as foreign in all organ transplants unless the donor is a twin.
- * c. The function of suppressor T cells is to suppress invading organisms.
- d. Each pathogen has its own unique antigen.
- e. The clonal selection theory holds that an activated B cell or T cell divides rapidly to produce a clone of immunologically identical cells that are specific for the antigen that selected them.

E   **32.** All of the cells involved in the immune response are
- a. leukocytes.
- b. erythrocytes.
- c. white blood cells.
- * d. both leukocytes and white blood cells.
- e. both erythrocytes and white blood cells.

E   **33.** The markers that identify "self" are actually
- a. genes.
- * b. proteins.
- c. lipids.
- d. small surface bumps.
- e. three letters of the alphabet.

M   **34.** The markers for each cell in a body are referred to by the letters
- * a. MHC.
- b. HTC.
- c. ADS.
- d. RSW.
- e. AKA.

LYMPHOCYTE BATTLEFIELDS

E   35. Which of the following is NOT included in the lymphatic system?
    a. tonsils
    b. lymphoid nodules
    c. thymus gland
  * d. liver
    e. spleen

E   36. Which of the following acts as "filters" in the lymph system?
    a. macrophages
  * b. lymph nodes
    c. complements
    d. immunoglobulins
    e. perforins

CELL-MEDIATED RESPONSES

M   37. Which cells are similar to cytotoxic cells in their mode of action?
    a. helper T
  * b. natural killer
    c. cytotoxic T
    d. memory
    e. B

M   38. Cell-mediated response
  * a. involves cytotoxic T cells.
    b. involves the action of antibodies to destroy invaders.
    c. acts only on extracellular clues.
    d. results in the production of clones of plasma cells.
    e. all of these

M   39. Which cells directly destroy body cells infected by viral or fungal parasites?
    a. helper T
    b. suppressor T
  * c. cytotoxic T
    d. memory
    e. B

M   40. Which of the following would be ignored in most instances by lymphocytes?
    a. cells coated with complement proteins
    b. cells with antigens on their surface
  * c. "self" cells with MHC markers
    d. cells with both antigen and self-MHC markers
    e. cells with damaged or mutant self-MHC markers

D   41. Which of the following statements is false?
    a. Cytotoxic T cells kill body cells that have been invaded by pathogens.
  * b. The cell-mediated response is ineffective against a pathogen that has already entered the cytoplasm of a body cell.
    c. Cytotoxic T cells are produced by the bone marrow but mature in the thymus gland.
    d. Cytotoxic T cells are unable to destroy free-floating viruses they encounter in the bloodstream.
    e. Cytotoxic T cells secrete perforins that are able to punch holes in infected cells.

ANTIBODY-MEDIATED RESPONSES

E   42. Antibodies are
  * a. proteins.
    b. steroids.
    c. polysaccharides.
    d. lipoproteins.
    e. all of these

E   43. Antibodies belong to a group of compounds known as
    a. self-recognizing compounds.
  * b. immunoglobulins.
    c. histosaccharides.
    d. antisteroids.
    e. virulent bases.

D   44. Which statement is NOT true?
    a. When an invading bacterium is destroyed by a macrophage, its antigens are preserved.
  * b. Antibodies attack and destroy invading antigens.
    c. Helper T cells recognize the major histocompatibility complex and antigens on the surface of macrophages.
    d. Self cells have major histocompatibility complex markers or antigens.
    e. Helper T cells secrete lymphokines, which help the cells of the immune system communicate with each other.

M   45. Effector (plasma) cells
    a. die within a week of production.
    b. manufacture and secrete antibodies.
    c. do not divide and form clones.
    d. develop from B cells.
  * e. all of these

M **46.** Which immunoglobulin is able to pass the placenta to protect the fetus from pathogens?
* a. IgG
  b. IgA
  c. IgD
  d. IgM
  e. IgE

M **47.** Most organ transplants fail because
  a. of poor vascular connection between host and donor tissue.
  b. the migrating leukocytes attack the tissue adjacent to the transplant.
* c. cytotoxic T cells enter the transplant through the connecting blood vessels and kill the individual transplant tissue.
  d. introduced tissues produce antibodies that cause a massive reaction.
  e. all of these

E **48.** Organ transplants are safest between
  a. two brothers.
  b. father and daughter.
  c. fraternal twins.
* d. identical twins.
  e. unrelated individuals.

IMMUNE SPECIFICITY AND MEMORY

E **49.** Antibodies are shaped like the letter
  a. C.
  b. E.
  c. H.
  d. K.
* e. Y.

E **50.** The antibody molecule consists of how many polypeptide chains, including light and heavy chains?
  a. 2
  b. 3
* c. 4
  d. 5
  e. 6

M **51.** Body cells have self-markers located
  a. in their nuclei.
  b. in the endoplasmic reticulum.
  c. in the mitochondria.
* d. on the plasma membrane.
  e. inside the Golgi bodies.

M **52.** Clones of B or T cells are
  a. being produced continually.
  b. interchangeable.
* c. produced only when their surface proteins recognize specific protein.
  d. known as memory cells.
  e. produced and mature in the bone marrow.

D **53.** The infinite variety of antibodies that can be generated by B cells is due to
  a. the infinite variety of genes in these cells.
* b. the shuffling of genes to produce an infinite variety of proteins.
  c. the recombination of genes due to crossing over.
  d. the infinite variety of genes in these cells and the recombination of genes due to crossing over.
  e. the infinite variety of genes in these cells, the recombination of genes due to crossing over, and the shuffling of genes to produce an infinite variety of proteins.

PRACTICAL APPLICATIONS OF IMMUNOLOGY

M **54.** The primary immune response
  a. is shorter in duration than a secondary response.
  b. is quicker than a secondary response.
  c. depends on random construction of appropriate antibodies.
  d. is the result of a reproduction of an appropriate lymphocyte resulting in a sensitive clone.
* e. both depends on random construction of appropriate antibodies and is the result of a reproduction of an appropriate lymphocyte resulting in a sensitive clone.

D   55. After a primary immune response
         a. a clone of sensitive lymphocytes is
            ready for any subsequent invasion of
            the same antigen.
         b. some of the clone cells will remain
            alive for decades.
         c. clone cells may be modified to attack
            new invaders.
         d. clone cells are continually reproduced to
            confer immunity against subsequent
            invasion.
      * e. a clone of sensitive lymphocytes is
            ready for any subsequent invasion of
            the same antigen and some of the clone
            cells will remain alive for decades.

M   56. Which of the following statements is false?
         a. Only B cells and their progeny make
            antibodies.
      * b. The primary immune response is faster
            and more complete than a secondary
            immune response.
         c. Virgin B cells already have antibodies
            but have not yet encountered an
            antigen.
         d. Macrophages will digest invading
            bacterial cells but do not destroy the
            antigens that eventually become
            mounted on the surface of the
            macrophages.
         e. Some B cell progeny differentiate into
            memory cells.

E   57. A vaccine may contain
         a. killed pathogen.
         b. weakened pathogen.
         c. noninfective fragments of a pathogen.
         d. full-strength pathogen.
      * e. any of these except full-strength
            pathogen.

M   58. Passive immunity can be obtained by
         a. having the disease.
         b. receiving a vaccination against the
            disease.
         c. receiving antibodies by injection.
         d. receiving antibodies from mother at
            birth.
      * e. either receiving antibodies by injection
            or receiving antibodies from mother at
            birth.

D   59. The purpose of a vaccine is to
         a. produce a mild case of the disease.
         b. stimulate the immune response.
         c. cause memory cells to be formed.
         d. stimulate the immune response and
            cause memory cells to be formed.
      * e. produce a mild case of the disease,
            stimulate the immune response, and
            cause memory cells to be formed.

M   60. Large amounts of antibodies produced for
         commercial use are called
      * a. monoclonal antibodies.
         b. myeloma cells.
         c. interferons.
         d. clones.
         e. hybridoma cells.

M   61. What is the name given to the chemicals
         which are produced by T cells and disrupt
         viral replication?
         a. lymphokines
         b. monoclonal antibodies
         c. vaccine
      * d. interferons
         e. complement

M   62. Interferon is a chemical produced by
         a. helper T cells.
         b. plasma cells.
         c. B cells.
         d. macrophages.
      * e. cells that have been invaded by a virus.

M   63. Monoclonal antibodies are produced by the
         fusion of
      * a. cancer cells and B cells.
         b. spleen cells and T cells.
         c. embryonic cells and plasma cells.
         d. thymus cells and helper T cells.
         e. marrow cells and B cells.

M   64. Monoclonal antibodies
         a. are formed by hybrid cells.
         b. are produced by cancer cells.
         c. are derived from spleen cells.
         d. produce identical antibodies and are all
            derived from the same parental cells.
      * e. all of these

ABNORMAL, HARMFUL, OR DEFICIENT IMMUNE RESPONSES

M   **65.** Whenever the body is reexposed to a sensitizing agent, the IgE antibodies cause

  *   a. the production of prostaglandins and histamine.
    b. the release of antihistamines.
    c. the suppression of the inflammatory response.
    d. the production of clonal cells.
    e. all of these

M   **66.** A person sensitive to bee stings may die minutes after a sting due to

    a. a collapse of the immune system.
    b. a clogging of the capillaries.
  *   c. a release of excessive fluids from the capillary beds.
    d. respiratory distress caused by excessive mucus.
    e. the extremely sharp rise in blood pressure.

M   **67.** Which of the following statements is false?

    a. Individuals are injected with antibodies in passive immunity.
    b. A genetically engineered virus is not as potentially dangerous as a weakened but intact pathogen.
    c. Allergies occur when the body makes a secondary immune response to a normally harmless substance.
  *   d. Allergies are a nuisance but are never dangerous or life-threatening.
    e. Allergies cause the secretion of mucus, prostaglandins, and histamines.

M   **68.** When the body's defenses turn against its own cells, the disorder is called

  *   a. an autoimmune response.
    b. anaphylactic shock.
    c. acquired immune deficiency syndrome.
    d. passive immunity.
    e. an inflammatory response.

E   **69.** Rheumatoid arthritis is

    a. sexually transmitted.
  *   b. an autoimmune disease.
    c. one of the diseases associated with AIDS.
    d. caused by a bacterial infection.
    e. preventable by vaccination.

E   **70.** The reason AIDS is so serious is that

    a. the excessive immune reaction leads to death.
    b. it is so highly contagious.
  *   c. it is fatal.
    d. it is caused by a retrovirus.
    e. many natural reservoirs may spread the disease at any time.

E   **71.** Of the following, AIDS is usually transferred by

    a. casual contact.
    b. food.
    c. water.
  *   d. sexual intercourse.
    e. insect bites.

M   **72.** The human immunodeficiency virus (HIV-1) primarily destroys which cells?

    a. B
    b. M
    c. T1
  *   d. helper T
    e. suppressor T

M   **73.** Which of the following statements is false?

    a. The virus that causes AIDS is the human immunodeficiency virus.
    b. The AIDS virus is a retrovirus.
    c. The AIDS virus attacks macrophages and helper T cells.
    d. Even though the AIDS virus does not have its own DNA, it causes the host to produce DNA that then becomes incorporated into host chromosomes.
  *   e. Antibodies to the AIDS virus can be detected immediately after infection occurs.

Matching Questions

D **74.** Choose the most appropriate answer for each.

1 ____ antigens

2 ____ B lymphocytes

3 ____ antibodies

4 ____ clone

5 ____ histamine

6 ____ cytotoxic T lymphocytes

7 ____ interleukins

8 ____ macrophages

9 ____ memory cells

10 ____ effector (plasma) cells

11 ____ retroviruses

12 ____ stem cells

13 ____ natural killer cells

A. cells that do not divide, die in less than a week, and secrete large quantities of antibodies

B. cells that directly destroy body cells already infected by viral or fungal parasites, as well as mutant and cancerous cells

C. lymphocytes that are held in reserve, circulate in the bloodstream, and enable a rapid response to subsequent encounters with the same type of invader

D. chemical that causes capillaries to become "leaky"

E. cells that are produced in the bone marrow, are never changed by the thymus; these cells produce antibodies

F. destroy cells coated with complement proteins

G. a class of proteins that help cells of the immune system communicate with each other

H. "big eaters" that alert other lymphocytes to the invasion of specific antigens

I. immature cells that may or may not be committed to develop into one of several mature cell types

J. a group of cells that are all produced asexually from one original parent cell

K. surface patterns of nonself molecules or particles

L. Y-shaped proteins that participate in immune responses

M. one of this group has been identified as the causative agent of AIDS

Answers:

1. K	2. E	3. L
4. J	5. D	6. B
7. G	8. H	9. C
10. A	11. M	12. I
13. F		

Classification Questions

Answer questions 75–79 in reference to the five types of white cells listed below:

a. macrophages

b. helper T cells

c. B cells

d. cytotoxic T cells

e. natural killer cells

D **75.** These cells do not need to encounter an antigen-MHC complex to take action.

M **76.** These cells scavenge dead cells and attack bacteria directly.

D **77.** These cells destroy cells infected by viruses.

D **78.** These cells recognize cell surface antigens and initiate the proliferation of lymphocytes.

M **79.** These cells produce antibodies.

Answers:

75. e	76. a	77. d
78. b	79. c	

Answer questions 80-84 in reference to the five items listed below:

    a. antigens
    b. antibodies
    c. helper T cells
    d. cytotoxic T cells
    e. memory B cells

E   **80.** These bind, as in a lock-and-key mechanism, to foreign proteins.

M   **81.** These produce immunoglobulins in response to the reinvasion by a virus.

D   **82.** These directly attack the foreign cells of an incompatible skin graft.

M   **83.** Rh^+ molecules in the body of an Rh^- woman represents these.

D   **84.** Bacteria and viruses in the blood are attacked by proteins produced by these.

Answers: 80. b   81. e   82. d
            83. a   84. e

Selecting the Exception

M   **85.** Four of the five answers listed below are barriers to invasion. Select the exception.
    a. intact skin
    b. mucous membrane
    c. gastric juices
  *   d. blood plasma
    e. lysozyme

D   **86.** Four of the five answers listed below are characteristic reactions of the complement system to invaders. Select the exception.
    a. causes an amplifying cascade of reactions to invaders
  *   b. triggers the secretion of histamines
    c. causes invading cells to lyse
    d. enhances the recognition of invaders by phagocytes
    e. creates gradients that attract phagocytes

D   **87.** Four of the five answers listed below are events of the inflammatory response. Select the exception.
    a. increases capillary permeability
    b. release of histamine
    c. blood vessels dilate
  *   d. temperature of the affected areas drops
    e. phagocytes migrate toward the affected area

E   **88.** Four of the five answers listed below are targets of the immune system. Select the exception.
    a. virus
  *   b. normal cells
    c. cancer cells
    d. bacteria
    e. debris and dead cells

CHAPTER 9

THE RESPIRATORY SYSTEM

Multiple-Choice Questions

OVERVIEW OF THE RESPIRATORY SYSTEM

M 1. What is the proper sequence in the flow of air in humans?
a. nasal cavities, larynx, pharynx, bronchi, trachea
b. nasal cavities, pharynx, bronchi, larynx, trachea
* c. nasal cavities, pharynx, larynx, trachea, bronchi
d. nasal cavities, larynx, pharynx, trachea, bronchi
e. nasal cavities, bronchi, larynx, trachea, pharynx

E 2. Which of the following is NOT a function of the nasal cavities?
a. Filter dust out of the incoming air.
b. Detect odors.
c. Warm the air.
* d. Oxygenate the blood.
e. Moisturize the air.

E 3. In describing the parts of the respiratory system, the word "septum" designates a partition in the
a. larynx.
* b. nasal cavities.
c. pharynx.
d. glottis.
e. lungs.

E 4. What is the name given to the respiratory ailment in which the bronchioles constrict severely?
a. pleurisy
b. emphysema
c. bronchitis
d. laryngitis
* e. asthma

M 5. The last mammalian structure that air moves through before the alveoli is the
a. larynx.
b. glottis.
* c. bronchioles.
d. trachea.
e. pharynx.

M 6. Food and drink are prevented from entering the respiratory passageways during swallowing by means of the
a. glottis.
b. pharynx.
* c. epiglottis.
d. larynx.
e. trachea.

M 7. When you swallow, the epiglottis covers the opening to the
a. pharynx.
b. esophagus.
* c. larynx.
d. bronchus.
e. alveoli.

E 8. The human vocal cords are located in the
a. glottis.
b. pharynx.
c. trachea.
* d. larynx.
e. bronchus.

M 9. The lungs move easily within their protective sacs due to
* a. intrapleural fluid.
b. leaking plasma.
c. blood.
d. mucus.
e. both leaking plasma and mucus.

M 10. In pleurisy,
a. some of the alveoli fill with fluid.
* b. the pleural membrane becomes inflamed and swollen and causes painful breathing.
c. the diaphragm develops muscular cramps.
d. the vagus nerve is irritated.
e. the intercostal muscles become inflamed and cause pain during deep breathing.

E   11. Actual exchange of gases in the lungs
          occurs in the
          a. bronchi.
     *   b. alveoli.
          c. bronchioles.
          d. tracheas.
          e. glottis.

A CLOSER LOOK AT GAS EXCHANGE

E   12. For an animal's surface to function in the
          integumentary exchange of gases it must
          a. be thin and soft.
          b. have a high number of blood vessels.
          c. have mucus or moist covering.
     *   d. all of these

E   13. Which vertebrate body system is most
          closely associated functionally with
          respiration?
          a. urinary
          b. digestive
          c. endocrine
     *   d. circulatory
          e. integumentary

E   14. The oxygen content of air is approximately
     *   a. 21 percent.
          b. 78 percent.
          c. 0.04 percent.
          d. 0.96 percent.
          e. 100 percent.

E   15. The most abundant gas in the earth's
          atmosphere is
          a. oxygen.
          b. water vapor.
          c. argon.
     *   d. nitrogen.
          e. carbon dioxide.

E   16. The concentration of nitrogen in the earth's
          atmosphere is approximately
     *   a. 78 percent.
          b. 66 percent.
          c. 50 percent.
          d. 33 percent.
          e. 20 percent.

E   17. The concentration of carbon dioxide in the
          earth's atmosphere is
          a. 0.004 percent.
     *   b. 0.04 percent.
          c. 0.4 percent.
          d. 4.0 percent.
          e. 40 percent.

E   18. The atmospheric pressure at sea level is
          a. 1,000mm Hg.
     *   b. 760mm Hg.
          c. 540mm Hg.
          d. 400mm Hg.
          e. 320mm Hg.

M   19. Exchange across a membrane requires
          a. moisture.
          b. transport proteins.
          c. pressure gradients.
     *   d. moisture and pressure gradients only.
          e. moisture, pressure gradients, and
              transport proteins..

M   20. The movement of both oxygen and carbon
          dioxide in the body is accomplished by
          a. exocytosis and endocytosis.
          b. bulk flow.
          c. osmosis.
     *   d. diffusion.
          e. facilitated diffusion.

E   21. Decompression sickness is caused by
          a. a rapid rise of carbon dioxide in the
              blood.
          b. lack of oxygen in the tissues.
     *   c. bubbles of nitrogen in the blood.
          d. glucose deficiency.
          e. descending too rapidly into deep water.

M   22. Hypoxia
          a. causes hyperventilation.
          b. may cause headaches, nausea, and
              lethargy.
          c. can lead to loss of consciousness and
              death.
          d. may be the result of changes in
              altitude.
     *   e. all of these

E   23. Carbon monoxide
          a. has a very low affinity or attraction to
              hemoglobin.
          b. is unlikely to be transported by the
              circulatory system.
          c. is not the cause of death of people who
              breathe excessive amounts of
              automobile exhausts.
     *   d. can arise from cigarette smoke.
          e. is not the cause of death of people who
              breathe excessive amounts of
              automobile exhausts but can arise from
              cigarette smoke.

D   **24.** Although carbon dioxide is normally carried by hemoglobin, the fact that carbon monoxide reduces hemoglobin's oxygen-carrying capacity would indicate that
     a. different hemoglobins carry different gases.
   * b. carbon monoxide competes with oxygen for the same binding sites on hemoglobin.
     c. binding of different gases is directed by a variety of enzymes.
     d. carbon dioxide is an abnormal gas.

BREATHING—CYCLIC REVERSALS IN AIR PRESSURE GRADIENTS

D   **25.** During inhalation,
     a. the pressure in the thoracic cavity is greater than the pressure within the lungs.
   * b. the pressure in the thoracic cavity is less than the pressure within the lungs.
     c. the diaphragm moves upward and becomes more curved.
     d. the chest cavity volume decreases.
     e. all of these

M   **26.** In humans ventilation is powered by
     a. the diaphragm.
     b. muscles attached to the ribs.
     c. the lungs themselves.
   * d. the diaphragm and rib muscles
     e. all of these

M   **27.** Which of the following is NOT found in lung tissue?
     a. blood capillaries
     b. alveolar sacs
     c. interstitial fluid
     d. connective tissue
   * e. muscle

M   **28.** During inhalation,
     a. the pressure in the thoracic cavity is greater than the pressure within the lungs.
   * b. the pressure in the thoracic cavity is less than the pressure within the lungs.
     c. the diaphragm moves upward and becomes more curved.
     d. the chest cavity volume decreases.
     e. all of these

E   **29.** The maximum amount of air that can be taken into the lungs in a single deep breath is the
   * a. vital capacity.
     b. tidal volume.
     c. pleural volume.
     d. alveolar volume.
     e. inspirational capacity.

E   **30.** The amount of air that moves into and out of the human lungs in each normal breath is termed
     a. inspirational capacity.
     b. reserve volume.
     c. pleural volume.
     d. alveolar volume.
   * e. tidal volume.

GAS EXCHANGE AND TRANSPORT

M   **31.** Oxygen moves from alveoli to the bloodstream
   * a. because the concentration of oxygen is greater in alveoli than in the blood.
     b. mainly due to the activity of carbonic anhydrase in the red blood cells.
     c. by using the assistance of carbaminohemoglobin.
     d. through active transport.
     e. all of these

M   **32.** Hemoglobin
     a. tends to give up oxygen in regions where partial pressure of oxygen exceeds that in the lungs.
     b. tends to hold onto oxygen when the pH of the blood drops.
     c. tends to release oxygen where the temperature is lower.
   * d. releases oxygen more readily in highly active tissues.
     e. all of these

M   **33.** Which statement is NOT true?
     a. Carbon dioxide is more soluble in fluid than in oxygen.
     b. Carbon dioxide diffuses more rapidly across the respiratory surface than does oxygen.
     c. The major muscle involved in breathing is the diaphragm.
   * d. Oxygen is carried primarily by blood plasma.
     e. Carbon dioxide is carried by the blood plasma.

M **34.** Hemoglobin gives up O_2 when
   * a. carbon dioxide concentrations are high.
      b. body temperature is lowered.
      c. pH values are high.
      d. CO_2 concentrations are low.
      e. all of these

M **35.** Most of the carbon dioxide produced by the body is transported to the lungs in
      a. a gaseous form.
      b. blood plasma.
      c. potassium carbonate ions.
   * d. bicarbonate ions.
      e. carbonic acid.

M **36.** The enzyme responsible for converting free carbon dioxide in the blood into forms in which it can be transported in the blood is
   * a. carbonic anhydrase.
      b. carboxypeptidase.
      c. carbonase.
      d. decarboxylase.
      e. dehydrogenase.

D **37.** Hemoglobin
      a. tends to release oxygen under warmer temperatures.
      b. picks up more oxygen the higher its partial pressure.
      c. picks up more oxygen when it is saturated.
      d. will give up oxygen when the partial pressure of oxygen is higher than it is in the lungs.
   * e. both tends to release oxygen under warmer temperatures and picks up more oxygen the higher its partial pressure.

D **38.** Which of the following statements is false?
      a. Hemoglobin functions as a buffer.
      b. Bicarbonate ions tend to diffuse out of red blood cells into the blood plasma.
      c. The movement of molecules is in different directions in the metabolically active tissues and the alveoli.
   * d. Carbonic anhydrase is an enzyme that promotes the formation of oxyhemoglobin.

M **39.** Carbonic anhydrase
      a. combines with water to form carbonic acid.
      b. dissociates into bicarbonate and hydrogen ions.
   * c. is normally found in red blood cells.
      d. is responsible for maintaining the high levels of carbon dioxide in the lungs and the low levels of carbon dioxide in the body tissues.
      e. all of these

CONTROLS OVER GAS EXCHANGE

M **40.** Which statement is true?
      a. Breathing rate and depth are completely under voluntary control.
      b. A person can commit suicide by holding one's breath.
      c. The contraction of the diaphragm and muscle of the rib cage are under the control of areas of the brain.
      d. There are chemoreceptors in the brain that monitor carbon dioxide content in the blood and control breathing.
   * e. The contraction of the diaphragm and muscle of the rib cage are under the control of areas of the brain; and There are chemoreceptors in the brain that monitor carbon dioxide content in the blood and control breathing.

M **41.** The rate and depth of breathing are governed by
      a. chemoreceptors in arterial walls.
      b. baroreceptors in the diaphragm.
      c. the partial pressure of O_2 in the atmosphere.
      d. a respiratory center in the brainstem.
   * e. all of these except "baroreceptors in the diaphragm."

M **42.** Exhaling into a paper bag and rebreathing the exhaled air would be expected to
      a. alert the brain.
      b. increase the breathing rate.
      c. stimulate the carotid bodies.
      d. increase the breathing rate and stimulate the carotid bodies.
   * e. alert the brain, increase the breathing rate, and stimulate the carotid bodies.

FOCUS ON YOUR HEALTH: TOBACCO AND OTHER THREATS TO THE RESPIRATORY SYSTEM

E   **43.** The cessation of smoking
   a. can reduce the risk of stillbirth.
   b. reduces the chances of cancer.
   c. reduces the chances of coronary disease.
   d. improves lung functioning.
 * e. all of these

M   **44.** Life-long nonsmokers live an average of
how much longer than those who, in their
mid-twenties, smoked two packs of
cigarettes a day?
   a. 6 months
   b. 1–2 years
   c. 3–5 years
 * d. 7–9 years
   e. over 12 years

E   **45.** Smoking has been shown to cause
   a. bronchitis.
   b. emphysema.
   c. lung cancer.
   d. coronary disease.
 * e. all of these

Matching Questions

D   **46.** Choose the one most appropriate answer for
each.

1 _____   alveoli

2 _____   bronchi

3 _____   diaphragm

4 _____   epiglottis

5 _____   intercostal rib muscles

6 _____   larynx

7 _____   pharynx

8 _____   trachea

   A. flexible windpipe reinforced with
cartridges

   B. contains two true vocal cords

   C. move the ribs

   D. throat cavity behind the mouth

   E. connect trachea to lungs

   F. flaplike structure that points upward
and allows air to enter trachea;
closed during swallowing

   G. contraction moves it downward

Answers: 1. H   2. E   3. G
             4. F   5. C   6. B
             7. D   8. A

Classification Questions

Answer questions 47-51 in reference to the five
components of respiratory systems listed below:

   a. pharynx
   b. larynx
   c. trachea
   d. bronchiole
   e. alveolus

E   **47.** This is the location of the voice box.

M   **48.** This is the last component of the human
lung that air flows into.

M   **49.** This is the site of gas exchange between the
air in the lungs and their blood supply.

E   **50.** Air moves from the nasal cavity into this
component.

M   **51.** Spent air moves from the bronchial tubes
back to this component.

Answers: 47. b   48. e   49. e
             50. a   51. c

Selecting the Exception

E   **52.** Four of the five answers listed below are
parts of the same body system. Select the
exception.
   a. trachea
 * b. esophagus
   c. alveoli
   d. bronchiole
   e. glottis

M   **53.** Four of the five answers listed below are
components of the human respiratory
system. Select the exception.
   a. thoracic cavity
   b. trachea
   c. diaphragm
 * d. spiracle
   e. larynx

D   54. Four of the five answers listed below are
            related by the same function. Select the
            exception.
                a. blood plasma
                b. carbaminohemoglobin
        *   c. oxyhemoglobin
                d. carbonic acid
                e. bicarbonate ions

D   55. Four of the five answers listed below are
            related by a similar function. Select the
            exception.
        *   a. intercostal muscles
                b. medulla oblongata
                c. aortic bodies
                d. reticular formation
                e. carotid bodies

CHAPTER 10
WATER-SALT BALANCE AND EXCRETION

Multiple-Choice Questions

THE CHALLENGE: SHIFTS IN EXTRACELLULAR FLUID

E   **1.** Extracellular fluid includes
     a. interstitial fluid.
     b. blood.
     c. lymph.
     d. blood and lymph only.
  *   e. interstitial fluid, blood, and lymph.

E   **2.** The most abundant waste product of metabolism is
  *   a. carbon dioxide.
     b. ammonia.
     c. urea.
     d. uric acid.
     e. water.

M   **3.** All but which of the following are significant routes for water loss from the body?
     a. excretion in urine
  *   b. sneezing
     c. sweating
     d. elimination in feces
     e. evaporation from respiratory surfaces

E   **4.** The process that normally exerts the greatest control over the water balance of an individual is
     a. sweating.
     b. elimination in feces.
  *   c. urinary excretion.
     d. evaporation through the skin.
     e. respiratory loss.

D   **5.** Which of the following does NOT dispose of a type of waste directly to the environment?
     a. digestive system
     b. respiratory system
     c. integumentary system
  *   d. circulatory system
     e. urinary system

D   **6.** Humans gain the least amounts of water by what route?
  *   a. metabolism
     b. ingestion of liquids
     c. ingestion of solids
     d. All of the above contribute equal amounts of water.

D   **7.** The most toxic substances routinely found in the blood are metabolites of
     a. carbohydrates.
  *   b. proteins.
     c. lipids.
     d. minerals.
     e. vitamins.

D   **8.** Which of the following solutes would NOT leave the vertebrate body under normal conditions?
     a. nutrients
     b. ammonia
     c. urea
     d. carbon dioxide
  *   e. hormones

THE URINARY SYSTEM

E   **9.** The subunit of a kidney that purifies blood and restores solute and water balance is called a
     a. glomerulus.
     b. loop of Henle.
  *   c. nephron.
     d. ureter.
     e. all of these

M   **10.** In the kidney, the collecting ducts from the nephrons empty immediately into the
     a. renal cortex.
     b. renal medulla.
  *   c. renal pelvis.
     d. ureter.
     e. urethra.

E 11. The last portion of the excretory system through which urine passes before it is eliminated from the body is the
   a. glomerulus.
   b. ureter.
 * c. urethra.
   d. bladder.
   e. rectum.

E 12. The functional unit of the kidney is the
   a. Bowman's capsule.
 * b. nephron.
   c. glomerulus.
   d. urinary bladder.

E 13. Filtration of the blood in the kidney takes place in the
   a. loop of Henle.
 * b. glomerulus.
   c. distal tubule.
   d. proximal tubule.
   e. all of these

M 14. Blood is delivered to each nephron by
   a. an efferent arteriole.
   b. peritubular capillaries.
   c. a renal capsule.
 * d. an afferent arteriole.
   e. proximal tubules.

D 15. After the blood leaves the glomerular capillaries, it next goes to the
   a. renal vein.
   b. renal artery.
 * c. peritubular capillaries.
   d. vena cava.
   e. heart.

HOW URINE FORMS

M 16. Filtration occurs in which section of mammalian nephrons?
 * a. glomerulus
   b. loops of Henle
   c. proximal tubules
   d. distal tubules
   e. peritubular capillaries

M 17. The process during which potassium and hydrogen ions, penicillin, and some toxic substances are put into the urine by active transport is called
 * a. tubular secretion.
   b. reabsorption.
   c. filtration.
   d. countercurrent multiplication.

M 18. Which of the following processes is under voluntary control?
   a. filtration
   b. reabsorption
 * c. urination
   d. secretion
   e. excretion

D 19. Kidney stones form in the
 * a. renal pelvis.
   b. ureters.
   c. urethra.
   d. urinary bladder.
   e. glomerulus.

E 20. Which of the following substances is NOT filtered from the bloodstream?
   a. water
 * b. plasma proteins
   c. urea
   d. glucose
   e. sodium

M 21. What is the name given to the fluid removed from the blood but not yet processed by the nephron tubules?
   a. urine
   b. water
   c. uretrial fluid
 * d. filtrate
   e. renal plasma

D 22. The process of filtration in the glomerulus is driven by
   a. active transport.
 * b. hydrostatic pressure.
   c. osmosis.
   d. dialysis.
   e. sodium-potassium pumps.

M 23. Reabsorption is the movement of water and solutes from the _____ to the _____.
   a. interstitial fluid; tubules
   b. glomerular capillaries; Bowman's capsule
   c. Bowman's capsule; nephron tubules
 * d. nephron tubules; capillaries
   e. glomerular capillaries; peritubular capillaries

E   **24.** In reabsorption,
 a. plasma proteins are returned to the blood.
 b. excess hydrogen ions are removed from the blood.
 c. excess water is passed on to the urine.
 * d. nutrients and salts are selectively returned to the blood.
 e. drugs and foreign substances are passed into the urine.

E   **25.** About what percent of the fluid removed from the blood is eventually returned to the blood?
 a. 59
 b. 90
 * c. 98
 d. 0.9
 e. 9

REABSORPTION OF WATER AND SODIUM

D   **26.** Most of the water and sodium is reabsorbed in the
 a. glomerulus.
 * b. proximal tubule.
 c. distal tubule.
 d. loop of Henle.
 e. collecting duct.

D   **27.** The reabsorption of solutes is the result of active transport of
 a. potassium.
 * b. sodium.
 c. carbonate.
 d. chloride.
 e. all of these

M   **28.** Water reabsorption into the capillaries associated with a nephron is achieved principally by
 a. bulk flow.
 * b. active transport and diffusion.
 c. countercurrent multiplication.
 d. phagocytosis.
 e. all of these

M   **29.** During reabsorption, sodium ions cross the proximal tubule walls into the interstitial fluid principally by means of
 a. bulk flow.
 * b. active transport.
 c. countercurrent multiplication.
 d. phagocytosis.
 e. all of these

M   **30.** Which of the following is actively transported in the proximal tubules of the kidney?
 a. bicarbonate ions
 * b. sodium ions
 c. chloride ions
 d. water

M   **31.** The longer this structure is, the greater is an animal's capacity to conserve water and to concentrate solutes for excretion in the urine.
 a. Bowman's capsule
 * b. loop of Henle
 c. proximal tubule
 d. ureter

D   **32.** Which of the following features would tend to promote water retention by the kidney?
 a. many nephridia
 * b. a long loop of Henle
 c. a long proximal tubule
 d. a short distal tubule
 e. a high filtration rate

E   **33.** A kidney machine removes solutes from the blood by means of
 a. osmosis.
 b. diffusion.
 * c. dialysis.
 d. active transport.
 e. bulk flow.

HORMONE ADJUSTMENTS OF REABSORPTION

M   **34.** The hormone that influences sodium reabsorption in the kidney is
 a. antidiuretic hormone.
 b. cortisone.
 * c. aldosterone.
 d. corticotropic hormone.
 e. adrenalin.

E   **35.** The hormone that controls the concentration of urine is
 a. insulin.
 b. glucagon.
 * c. antidiuretic hormone.
 d. thyroxine.
 e. epinephrine.

M **36.** Which of the following is a trigger for the other actions described?
   a. Posterior pituitary secretes ADH.
   b. Solute concentration in the extracellular fluid rises above a set point.
   c. Distal tubules of the nephrons and the collecting ducts become more permeable to water.
   * d. Extracellular fluid volume is reduced.
   e. A small amount of concentrated urine is excreted.

D **37.** A rise in sodium levels and extracellular volume leads to a rise in blood pressure. As a result
   a. renin levels rise, but aldosterone levels fall.
   b. renin levels fall, but aldosterone levels rise.
   c. renin and aldosterone levels rise.
   * d. renin and aldosterone levels drop.

M **38.** The hormonal control over excretion occurs in the
   a. Bowman's capsule.
   b. proximal tubule.
   * c. distal tubule.
   d. loop of Henle.
   e. urinary bladder.

M **39.** The antidiuretic hormone
   a. promotes processes that lead to an increase in the volume of urine.
   * b. promotes processes that lead to a decrease in the volume of urine.
   c. acts on the proximal tubules of nephrons in the kidney.
   d. is produced by the adrenal cortex.
   e. all of these

M **40.** In mammals which of the following governs both the thirst mechanism and the hormonal action that affects the amount of water and solutes excreted in the urine?
   a. adrenal cortex
   b. adrenal medulla
   c. anterior pituitary
   * d. hypothalamus
   e. none of these

M **41.** Which of the following processes occurs first in the adjustment of body fluid volume?
   a. An inactive protein is converted into angiotensin.
   b. Distal tubules and collecting ducts reabsorb sodium faster.
   c. Target cells secrete aldosterone.
   * d. Sense receptors in walls of blood vessels and the heart reveal a drop in extracellular fluid level.
   e. Renin is secreted by the juxtaglomerular apparatus.

D **42.** When the body has excess sodium, which of the following does NOT happen?
   a. More sodium is excreted.
   * b. More sodium is reabsorbed.
   c. Edema (swelling) occurs.
   d. Blood pressure rises.
   e. Aldosterone secretion is inhibited.

D **43.** Ethanol (drinking alcohol) is an inhibitor of ADH. Therefore, a person consuming a couple of mixed drinks should excrete
   a. less water because ADH promotes reabsorption.
   b. the alcohol because ADH cannot degrade it.
   c. ketone bodies formed from the alcohol.
   * d. more water because ADH normally promotes reabsorption.
   e. more water plus the alcohol due to the ADH inhibition.

E **44.** Which of the following does NOT belong with the others?
   a. angiotensins
   b. ADH
   c. aldosterone
   d. renin
   * e. insulin

E **45.** In humans, the thirst center is located in the
   a. adrenal cortex.
   * b. hypothalamus.
   c. anterior pituitary.
   d. glomerulus.
   e. stomach.

THE ACID-BASE BALANCE

D 46. Which of the following does NOT influence the pH of the blood and extracellular fluids?
    a. respiration
    b. blood proteins
    c. bicarbonate ions
*   d. filtration by glomerulus
    e. phosphate and ammonia ions

D 47. The urinary system helps to maintain the extracellular fluid pH by
    a. synthesizing buffers.
    b. retaining carbon dioxide in the filtrate.
*   c. excreting hydrogen ions as water.
    d. combining hydrogen ions with urea.

E 48. The normal pH of the extracellular fluid of the human body is about
    a. 3.2
    b. 8.5
    c. 1.1
*   d. 7.4
    e. 6.3

M 49. Hydrogen ions can be neutralized temporarily by which of the following?
*   a. bicarbonate
    b. ADH
    c. renin
    d. urea
    e. filtration

MAINTAINING THE BODY'S CORE TEMPERATURE

E 50. The usual upper temperature limit before proteins are denatured is
    a. 35°C.
*   b. 41°C.
    c. 45°C.
    d. 50°C.
    e. 55°C.

M 51. The rate of a chemical reaction is cut in half for a drop of every
    a. 5°F.
*   b. 10°F.
    c. 15°F.
    d. 20°F.
    e. 25°F.

M 52. Which of the following is NOT a response to low temperature that increases the chance for survival?
*   a. hypothermia
    b. shivering
    c. production of brown fat
    d. pilomotor response
    e. increased metabolism

M 53. The primary thermostat of the body is located in the
    a. heart.
*   b. hypothalamus.
    c. medulla oblongata.
    d. cerebellum.
    e. thyroid gland.

M 54. Which of the following is NOT part of the initial response to cold temperature?
*   a. opening of peripheral blood vessels
    b. shivering
    c. increased respiration
    d. shunting of the blood to the core regions of the body
    e. increased metabolism

M 55. In mammals, which of the following is the seat of temperature control?
    a. adrenal cortex
    b. adrenal medulla
    c. thymus
*   d. hypothalamus
    e. heart

E 56. Responses to heat stress include
    a. reduction in muscle contraction.
    b. increased sweating.
    c. dilation of peripheral blood vessels.
    d. loss of salts and liquids.
*   e. all of these

M 57. Which of the following statements about fever is NOT correct?
    a. Prostaglandins influence the hypothalamus.
    b. Fever is the result of a resetting of the body's "thermostat."
    c. Low fevers enhance the body's defense mechanisms.
*   d. When a person feels "chills", the body core temperature is decreasing.
    e. Aspirin exerts its fever reducing effects by interfering with prostaglandins.

Matching Questions

D **58.** Choose the one most appropriate answer for each.

1 ____ nephron

2 ____ aldosterone

3 ____ renin

4 ____ ADH

5 ____ glomerular filtration

6 ____ pilomotor response

7 ____ tubular reabsorption

A. flow of protein-free fluid from capillaries into Bowman's capsule

B. substance that acts on the adrenal glands to release aldosterone

C. a long, slender tubular unit in the vertebrate kidney that forms urine

D. secreted by adrenal glands; influences sodium reabsorption

E. passive transport of water; active and passive transport of solutes out of the nephron into peritubular capillaries

F. released from the posterior lobe of the pituitary in response to hypothalamic signals

G. smooth muscles cause body hairs to become erect

Answers:

1. C	2. D	3. B
4. F	5. A	6. G
7. E		

Classification Questions

Answer questions 59-63 in reference to the four regions of a nephron listed below:

a. Bowman's capsule
b. proximal tubule
c. descending portion of loop of Henle
d. distal tubule

M **59.** This portion immediately precedes the loop of Henle.

M **60.** Filtration of the blood occurs in association with this structure.

D **61.** Antibiotics are secreted from this structure.

M **62.** Permeability to water is regulated by antidiuretic hormone in this structure.

E **63.** The glomerular capillaries are intimately associated with this structure.

Answers:

59. b	60. a	61. d
62. d	63. a	

Selecting the Exception

D **64.** Four of the five answers listed below are potentially toxic waste products of metabolism. Select the exception.

a. urea
* b. water
c. uric acid
d. carbon dioxide
e. ammonia

M **65.** Four of the five answers listed below are parts of the same organ. Select the exception.

a. distal tubule
b. loop of Henle
* c. ureter
d. proximal tubule
e. Bowman's capsule

M **66.** Four of the five answers listed below are functions of the nephron. Select the exception.

a. filtration
* b. dilution
c. excretion
d. reabsorption
e. secretion

D **67.** Four of the five answers listed below are the result of ADH (antidiuretic hormone) secretion. Select the exception.
   a. water is reabsorbed in the distal tubule
   b. fluid volume of blood increases
 * c. rise in solute concentration in blood
   d. distal tubule and collecting duct become more permeable to water
   e. solutes concentration decreases

M **68.** Four of the five answers listed below are responses to cold temperatures. Select the exception.
   a. vasoconstriction of peripheral blood vessels
   b. uncontrolled muscular contraction
   c. thermoreceptors send messages to the hypothalamus
 * d. enhanced secretion of sweat glands
   e. increased metabolic reactions

M **69.** Four of the five answers listed below are responses of the body to heat. Select the exception.
   a. increased sweating
 * b. increased retention of blood on the core region
   c. decreased rate of muscle activity
   d. increased water loss
   e. dilation of peripheral blood vessels

D **70.** Four of the five answers listed below are responses to the loss of body heat. Select the exception.
 * a. ventricular fibrillation
   b. blood routed to deeper tissues
   c. constriction of peripheral blood vessels
   d. increased respiration
   e. increased metabolic rates

CHAPTER 11

THE NERVOUS SYSTEM

Multiple-Choice Questions

NEURONS—THE COMMUNICATION SPECIALISTS

E   1. The principal job of the human nervous system is to
*   a. facilitate communication among the body systems.
      b. store information.
      c. replace or repair damaged tissues.
      d. provide for defense against pathogens.
      e. rid the body of metabolic wastes.

E   2. The basic unit of the nervous system is
*   a. the neuron.
      b. neuroglia.
      c. the brain.
      d. a nerve.
      e. a nerve impulse.

M   3. Which of the following is NOT true concerning sensory neurons?
      a. They have receptor regions for detection of stimuli.
*   b. They lie in the pathway between the interneurons and motor neurons.
      c. They relay information to the spinal cord.
      d. They are part of a reflex arc.
      e. They are one of three types of neurons.

E   4. The single long process that extends from a typical motor nerve cell is the
*   a. axon.
      b. neuron.
      c. synapse.
      d. dendrite.

D   5. Within a single neuron, the direction an impulse follows is
      a. dendrite >>> axon >>> cell body.
      b. axon >>> dendrite >>>cell body.
*   c. dendrite >>> cell body >>> axon.
      d. cell body >>> dendrite >>> axon.
      e. cell body >>> axon >>> dendrite.

D   6. When an impulse passes from one neuron to the next, it
      a. is passed directly from dendrite to axon.
      b. passes from axon to cell body to dendrite.
      c. can bypass the cell bodies of both.
*   d. passes from axon to dendrite.
      e. undergoes repolarization.

M   7. Neuroglial cells
      a. metabolically support other neurons.
      b. form sheaths around neurons and control the rate of impulse transmission.
      c. form more than half of the volume of the brain.
      d. provide physical support.
*   e. all of these

M   8. Functionally speaking, a nerve impulse is
      a. a flow of electrons along the outside of the plasma membrane of a neuron.
      b. the movement of cytoplasmic elements through the core of the neuron.
*   c. a series of changes in membrane potentials.
      d. a lengthening and shortening of the membrane extensions of an individual neuron.
      e. a change in the metabolic rate within a neuron.

M   9. When a neuron is at rest
      a. there is a voltage difference across the membrane of about -70 millivolts.
      b. the interior is negatively charged.
      c. it is not responding to a stimulus.
      d. the fluid outside the membrane has more sodium and less potassium than the cytoplasm.
*   e. all of these

E   10. The resting potential of a neuron is approximately minus
      a. 70 microvolts.
*   b. 70 millivolts.
      c. 70 volts.
      d. 70 electrovolts.
      e. 70 megavolts.

M   **11.** The membrane-bound enzyme system that restores and maintains the resting membrane potential is which of the following pumps?
       a. sodium-phosphorus
*   b. sodium-potassium
       c. sodium-chlorine
       d. phosphorus-calcium
       e. phosphorus-chlorine

M   **12.** At rest, a nerve cell has a high concentration of _____ inside and a high concentration of _____ outside.
       a. acetylcholine; chlorine
       b. sodium; potassium
*   c. potassium; sodium
       d. calcium; phosphorous
       e. phosphorus; calcium

D   **13.** Active transport
       a. helps establish the resting potential of a neuron.
       b. counters the process of diffusion.
       c. allows transport of atoms across the plasma membrane of the neuron against the concentration gradient.
       d. both helps to establish the resting potential of a neuron and counters the process of diffusion.
*   e. helps to establish the resting potential of a neuron, counters the process of diffusion, and allows transport of atoms across the plasma membrane of the neuron against the concentration gradient.

D   **14.** Which of the following concerning transport proteins in the neuron membrane is true?
       a. Only sodium is transported.
       b. Only potassium is transported.
       c. Movement is in response to concentration gradients.
*   d. Energy moves ions against the concentration gradient.

A CLOSER LOOK AT ACTION POTENTIALS

D   **15.** Which of the following is the first response a neuron makes to a stimulus?
*   a. Sodium ions enter the cell.
       b. Sodium ions leave the cell.
       c. Potassium ions enter the cell.
       d. Potassium ions leave the cell.

D   **16.** For sodium to accumulate rapidly in a neuron,
*   a. a stimulus above the threshold must open sodium gates in an accelerating manner.
       b. the wave of repolarization must occur to reestablish a resting potential.
       c. there must be a dramatic increase in the negative charge of the cytoplasm.
       d. a voltage surge must cause the sodium gates to close.
       e. the potassium gates must open first.

M   **17.** Disturbances in sensory neurons will result in an action potential if the
       a. stimulus is graded.
       b. stimulus remains local.
*   c. graded stimulus reaches a trigger zone.
       d. localized stimuli do not spread too far.
       e. stimuli become downgraded to localized ones.

D   **18.** Which of the following occurs first during an action potential?
       a. Many sodium ions flow into the neuron.
       b. Voltage-gated sodium channels open.
*   c. A local disturbance triggers the resting voltage to exceed the threshold level.
       d. The interior of the neuron becomes positive.
       e. The interior of the neuron becomes negative.

D   **19.** Which of the following terms most accurately describes the cellular activity associated with the actual passage of a nerve impulse?
       a. electrical discharge
       b. action of sodium-potassium pump
*   c. wave of depolarization
       d. repolarization
       e. active transport of ions

D   **20.** What happens immediately following the closing of the sodium gates during an action potential?
       a. Sodium ions enter the cell.
       b. Sodium ions leave the cell.
       c. Potassium ions enter the cell.
*   d. Potassium ions leave the cell.

D   21. An action potential is brought about by
   a. a sudden membrane impermeability.
   b. the movement of negatively charged proteins through the neuronal membrane.
   c. the movement of lipoproteins to the outer membrane.
   * d. a local change in membrane permeability caused by a greater-than-threshold stimulus.
   e. all of these

D   22. During the passage of a nerve impulse
   a. sodium ions pass through gated channels.
   b. positive feedback causes more sodium ions to enter the cell.
   c. the interior of the cell becomes positive.
   d. changing voltage increases the number of open gates.
   * e. all of these

D   23. The phrase "all or nothing," used in conjunction with discussion about an action potential, means that
   a. a resting membrane potential has been received by the cell.
   * b. nothing can stop the action potential once the threshold is reached.
   c. the membrane either achieves total equilibrium or remains as far from equilibrium as possible.
   d. propagation along the neuron is saltatory.
   e. none of these

D   24. Once a threshold is reached,
   a. the number of sodium gates that open depends upon the strength of a stimulus.
   b. a graded local potential will be unable to spread to a trigger zone of the nerve membrane.
   c. the resting potential of a neuron is restored.
   d. the potassium channels in the input zone open.
   * e. the opening of sodium gates and the accompanying flow of sodium ions is an example of positive feedback.

D   25. The occurrence of an action potential can best be compared to a
   * a. switch to turn a lamp on and off.
   b. volume control on a stereo.
   c. door to the classroom.
   d. room light dimmer switch.

D   26. Which statement is false?
   a. A nerve will not fire unless a stimulus exceeds the threshold.
   b. An action potential is an all-or-nothing event.
   * c. An action potential continues indefinitely until a quenching signal is released.
   d. An action potential is self-propagating.
   e. An action potential transmission depends on activities at the membrane.

M   27. The recovery time from the passage of a nerve impulse is called the
   a. polarized response.
   b. wave of depolarization.
   c. action potential propagation.
   * d. refractory period.
   e. saltatory period.

D   28. During the recovery period between action potentials,
   a. the threshold value is increased.
   b. the threshold value is reduced.
   * c. the sodium gates are shut and the potassium gates are opened.
   d. both the sodium and the potassium gates are shut.
   e. the nerve is said to be at the resting potential.

CHEMICAL SYNAPSES

M   29. Transmitter substances
   a. are expelled from the presynaptic cells.
   b. tend to destroy acetylcholine.
   c. enter the presynaptic cell to continue the passage of the impulse.
   d. interact with membrane receptors of the postsynaptic cells.
   * e. are expelled from the presynaptic cells and interact with membrane receptors of the postsynaptic cells.

E   **30.** Which is a junction between two neurons?
- a. Schwann cell
- * b. chemical synapse
- c. node
- d. sodium gate
- e. all of these

M   **31.** The operation of a synapse
- a. results from the passage of an electrical charge across the gap.
- * b. involves a transmitter substance from vesicles in the presynaptic neuron that acts upon a receptor site in the postsynaptic neuron.
- c. occurs only between two nerves.
- d. is limited only by the action of acetylcholinesterase.

D   **32.** The presynaptic neuron and postsynaptic neuron do not directly contact each other because
- a. one would inhibit the actions of the other.
- b. they never grow to sufficient length.
- c. the synaptic vesicles keep them apart.
- * d. this would cause continuous impulse transmission.
- e. acetylcholine prevents this action.

M   **33.** What bridges the gap between a sending and a receiving neuron?
- a. threshold value
- b. action potential
- * c. transmitter substance
- d. neurohormone
- e. all of these

M   **34.** Transmitter substances
- a. include acetylcholine.
- b. change the permeability of postsynaptic cells.
- c. may be excitatory or stimulatory.
- d. may participate in synaptic integration.
- * e. all of these

M   **35.** Endorphins are
- a. neuromodulators.
- b. stimulators of brain and nervous activity.
- c. inhibitors of pain perception.
- d. neuromodulators and stimulators of brain and nervous activity.
- * e. neuromodulators and inhibitors of pain perception.

D   **36.** Organophosphate insecticides kill by inhibiting acetylcholinesterase, an enzyme that degrades acetylcholine. What effect does this have?
- a. It allows continuous volleys of impulses.
- b. Control of vital organs is disrupted.
- c. "Start/stop" signals for breathing and heartbeat are not possible.
- d. Control of vital organs is disrupted; and "Start/stop" signals for breathing and heartbeat are not possible.
- * e. It allows continuous volleys of impulses; Control of vital organs is disrupted; and "Start/stop" signals for breathing and heartbeat are not possible.

D   **37.** At an inhibitory synapse
- a. no transmitter substances are released by the sending cell.
- * b. a transmitter substance produces changes in the receiving cell that drive the membrane potential away from threshold.
- c. no transmitter substance can bind to the receiving cell.
- d. a transmitter substance produces changes in the receiving cell that drive the membrane potential closer to threshold.

D   **38.** An excitatory postsynaptic potential
- a. is only one of several types of graded potential.
- b. has a hyperpolarizing effect.
- c. will drive the membrane away from its threshold.
- * d. is summed with an inhibitory postsynaptic potential at the input zone of a neuron in a process known as synaptic integration.
- e. none of these

M   **39.** Synaptic integration means
- a. all positive or excitatory stimuli are added together.
- b. the positive and negative ions neutralize each other.
- * c. excitatory and inhibitory signals are combined in a neuron.
- d. the adjacent neurons interact so that excitatory and inhibitory stimuli cancel each other out.
- e. all of these

INFORMATION PATHWAYS

M **40.** The myelin sheath
    a. is formed by the Schwann cell.
    b. speeds up the transmission of impulses.
    c. does not surround all nerves.
    d. extends from node to node.
    * e. all of these

M **41.** The spaces that separate adjacent Schwann cells are called
    a. neuroglia.
    b. myelin sheaths.
    * c. nodes.
    d. dendrites.
    e. synapses.

M **42.** Saltatory ("jumping") conduction
    a. occurs only in the central nervous system.
    b. is a quicker type of nerve conduction.
    c. occurs between nerves and muscles.
    d. involves the movement of impulses from node to node.
    * e. involves both a quicker type of nerve conduction and the movement of impulses from node to node.

D **43.** A deterioration in the myelin sheaths of motor axons to the lower leg would be expected to
    a. remove the restraints to ion movement and speed up impulse transmission.
    b. cause immobility of the leg due to cessation of impulses to leg muscles.
    * c. slow the rate of transmission and cause lack of motor control.
    d. have little effect because the sheaths are for insulation only.

M **44.** Which statement is false?
    a. Reflexes are the simplest of all nervous reactions.
    b. The nervous system required sense organs before organisms could perceive their environment.
    * c. Motor neurons lead toward the brain or central nervous system.
    d. Reflex actions are stereotyped and repeatable.
    e. All of the above statements are true.

E **45.** The simplest nerve pathway
    a. is located in the midbrain.
    * b. is the reflex pathway.
    c. is found in the lower part of the brain.
    d. is found in the autonomic nervous system.
    e. is in the flow of information from a sense receptor to the brain.

M **46.** In the knee-jerk reflex arc, the synapse between a sensory neuron and a motor neuron occurs where?
    a. in the brain
    b. between the receptor and the spinal cord
    c. within the muscle of the leg
    * d. within the spinal cord
    e. in the nerve leading to the spinal cord

M **47.** One example of a simple reflex involves the
    * a. contraction of a muscle when it is stretched.
    b. conscious message to move part of the body.
    c. receptor, the brain, and the effector.
    d. muscle action in a salute when a noncommissioned serviceman sees an officer.
    e. contraction of an antagonistic muscle when its opposite muscle relaxes.

M **48.** The stretch reflex
    a. is an adaptation that enables humans to stand upright.
    b. is activated by stretch-sensitive receptors inside the muscle spindles.
    c. is a simple, stereotyped, and repeatable motor action.
    d. is elicited by a sensory stimulus.
    * e. all of these

M **49.** The sequence of a simple reaction to a stimulus is
    * a. sense organ, sensory neuron, association neuron, motor neuron, effector.
    b. sense organ, sensory neuron, motor neuron, association neuron, effector.
    c. sense organ, motor neuron, sensory neuron, association neuron, effector.
    d. sense organ, motor neuron, association neuron, sensory neuron, effector.

THE NERVOUS SYSTEM: AN OVERVIEW

E   **50.** The two main divisions of the nervous system are
    a. somatic and autonomic.
    b. sensory and motor.
  * c. central and peripheral.
    d. sympathetic and parasympathetic.
    e. skeletal and visceral.

E   **51.** Clusters of cell bodies of neurons outside the central nervous system are known as
    a. nerve cords.
  * b. ganglia.
    c. a plexus.
    d. notochords.
    e. nerves.

M   **52.** The two MAJOR divisions of the vertebral nervous system are the
    a. autonomic and peripheral systems.
    b. sympathetic and parasympathetic systems.
    c. cranial and spinal nerves.
  * d. central and peripheral nervous systems.
    e. brain and spinal cord.

M   **53.** Which of the following would NOT be a part of the central nervous system?
    a. brain
    b. cerebellum
    c. medulla
  * d. spinal nerves
    e. neuroglia cells

M   **54.** Which of the following statements concerning the peripheral nervous system is false?
    a. Spinal nerves lead to and from the spinal cord.
    b. There are 31 pairs of spinal nerves.
  * c. Cranial nerves lead from the brain directly to the spinal cord.
    d. Some nerves carry only sensory information.
    e. Some nerves are both sensory and motor.

M   **55.** By definition, a nerve is
  * a. a bundle of axons.
    b. a single extension of a neuron.
    c. the same as a neuron within the central nervous system.
    d. a dendrite.
    e. a fiber more than 10 inches in length.

THE MAJOR EXPRESSWAYS

M   **56.** The two principal divisions of the peripheral nervous system are the
  * a. somatic and autonomic systems.
    b. sympathetic and parasympathetic systems.
    c. peripheral and central systems.
    d. afferent and autonomic systems.
    e. cranial and skeletal nerves.

E   **57.** All nerves that lead away from the central nervous system are
  * a. efferent nerves.
    b. sensory nerves.
    c. afferent nerves.
    d. spinal nerves.
    e. peripheral.

E   **58.** The autonomic subdivision consists specifically of
    a. central and peripheral nerves.
  * b. parasympathetic and sympathetic.
    c. somatic and involuntary.
    d. brain and spinal cord.
    e. spinal and cranial nerves.

M   **59.** The autonomic subdivision of the vertebrate nervous system would innervate all but which of the following?
    a. intestinal muscles
  * b. skeletal muscles
    c. heart
    d. pancreas
    e. liver

M   **60.** Which nerves generally dominate internal events when environmental conditions permit normal body functioning?
    a. ganglia
    b. pacemaker
    c. sympathetic
  * d. parasympathetic
    e. all of these

D   61. Which statement is true?
    a. Both the parasympathetic and sympathetic nervous systems send nerves to all organs.
    b. The sympathetic nervous system that supplies an organ will also provide parasympathetic nerves to it.
    c. The sympathetic branch can have either excitatory or inhibitory effects depending on effects from the environment.
  * d. The sympathetic branch of the autonomic system usually speeds up the activities of the body.
    e. The parasympathetic system usually speeds up the activities of the body.

E   62. Signals from the parasympathetic nervous system cause which of the following?
    a. rise in blood pressure
    b. increase in pulse rate
  * c. increase in digestive system movements
    d. rise in blood sugar level
    e. rise in metabolic rate

M   63. The parasympathetic nervous system includes the
    a. cranial and thoracic nerves.
    b. thoracic and lumbar nerves.
    c. lumbar and sacral nerves.
    d. cervical and lumbar nerves.
  * e. cranial and sacral nerves.

M   64. The sympathetic nervous system includes the
    a. cranial and thoracic nerves.
  * b. thoracic and lumbar nerves.
    c. lumbar and sacral nerves.
    d. cervical and lumbar nerves.
    e. cranial and sacral nerves.

M   65. Activation of the sympathetic nervous system
  * a. causes the pupils of the eye to dilate.
    b. increases the flow of watery saliva.
    c. stimulates peristaltic contractions of the intestinal system.
    d. slows heartbeat and lowers blood pressure.
    e. allows the body to relax rather than prepare for fight or flight.

D   66. The word that best describes the interaction of sympathetic and parasympathetic systems is
  * a. antagonistic.
    b. cooperation.
    c. overriding.
    d. subversive.
    e. ineffective.

M   67. The part of the central nervous system that is composed of parts that are antagonistic to each other is the
  * a. autonomic nervous system composed of the sympathetic and parasympathetic subsystems.
    b. central nervous system composed of the brain and spinal cord.
    c. peripheral nervous system composed of the cranial and spinal nervous system.
    d. none of these; the muscular system is the only system with antagonistic subsets.

E   68. Interneurons are found in the
    a. dorsal root.
  * b. spinal cord.
    c. sensory neurons.
    d. motor neurons.
    e. autonomic nervous system.

M   69. The ascending and descending nerve tracts
    a. are found in the gray matter of the spinal cord.
    b. are found in the white matter of the spinal cord.
    c. are covered with myelin sheaths.
    d. are both sensory and motor.
  * e. all except "are found in the gray matter of the spinal cord" are correct

D   70. All but which of the following are true of the spinal cord?
    a. It contains connections between sensory and motor neurons.
    b. It is a reflex center.
  * c. Nerve tracts to brain run through the gray matter.
    d. It is covered with meninges.
    e. It is enclosed by vertebrae.

E   71. Areas of the spinal cord appear glistening white because of
    a. naked dendrites.
    b. cell bodies.
    c. neuroglia cells.
    d. lack of meninges.
  * e. myelin sheaths.

THE BRAIN

E 72. The midbrain includes the
    a. thalamus.
    b. pineal gland.
    * c. tectum.
    d. medulla.
    e. olfactory lobes.

E 73. The hindbrain includes the
    a. thalamus.
    b. pineal gland.
    c. cerebellum.
    d. medulla.
    * e. both cerebellum and medulla.

E 74. The pituitary gland is controlled by the
    a. pineal gland.
    b. medulla.
    * c. hypothalamus.
    d. thalamus.
    e. cerebrum.

E 75. The center of consciousness and intelligence is the
    a. medulla.
    b. thalamus.
    c. pons.
    d. cerebellum.
    * e. cerebrum.

E 76. Which structure forms the roof of the midbrain where visual and auditory signals are integrated?
    a. ventricles
    b. meninges
    * c. tectum
    d. olfactory and optic bulbs
    e. pineal gland

E 77. The part of the brain that connects one brain center with another is the
    a. cerebrum.
    * b. pons.
    c. cerebellum.
    d. fissure of Rolando.
    e. hypothalamus.

E 78. The part of the brain that deals with the basic drives such as hunger, sex, and thirst is the
    a. cerebrum.
    b. pons.
    c. cerebellum.
    d. thalamus.
    * e. hypothalamus.

E 79. The major relay center of the brain is the
    a. cerebrum.
    b. olfactory area.
    c. cerebellum.
    * d. thalamus.
    e. hypothalamus.

E 80. The center for balance and coordination is the
    a. cerebrum.
    b. pons.
    * c. cerebellum.
    d. thalamus.
    e. hypothalamus.

E 81. The part of the brain that controls the basic responses necessary to maintain life processes (breathing, heartbeat) is the
    * a. medulla.
    b. corpus callosum.
    c. fissure of Rolando.
    d. cerebellum.
    e. cerebral cortex.

D 82. Destruction of the motor areas in the left cerebral cortex results in the loss of
    a. sensation on the right side of the body.
    b. sensation on the left side of the body.
    c. voluntary movement on the left side of the body.
    * d. voluntary movement on the right side of the body.

E 83. Which part of the mammalian brain is disproportionately larger than the corresponding part of a fish brain?
    a. medulla
    b. thalamus
    c. pons
    d. cerebellum
    * e. cerebrum

E 84. The protective covering of the brain is the
    a. ventricles.
    * b. meninges.
    c. tectum.
    d. olfactory and optic bulbs.
    e. pineal gland.

E 85. The chambers of the brain are the
    * a. ventricles.
    b. meninges.
    c. tectum.
    d. olfactory and optic bulbs.
    e. pineal gland.

E   **86.** The gray matter of the brain is associated with the

    *   a. cerebral cortex.
       b. pons.
       c. optic chiasm.
       d. corpus callosum.
       e. thalamus.

D   **87.** The blood-brain barrier consists of

       a. membranes surrounding the neurons of the brain.
    *   b. endothelial cells of the brain capillaries.
       c. vesicles of cerebrospinal fluid.
       d. the cerebral cortex.
       e. the meninges.

E   **88.** The cerebral hemispheres communicate with each other by means of the

       a. cerebral cortex.
       b. corpora cardiaca.
       c. corporothalamus.
    *   d. corpus callosum.
       e. corpus allata.

D   **89.** If the motor cortex on the right side of the brain is destroyed by a stroke, what would be impaired?

       a. movement on both sides of the body
       b. reception of sensory information from the left side of the body
       c. movement by the right side of the body
    *   d. movement by the left side of the body
       e. all of these

M   **90.** To produce a split-brain individual, an operation would need to cut the

    *   a. corpus callosum.
       b. reticular formation.
       c. hypothalamus.
       d. fissure of Rolando.
       e. pons.

M   **91.** The left hemisphere of the brain is responsible for

       a. music.
       b. artistic ability and spatial relationships.
       c. mathematics.
    *   d. language skills.
       e. abstract abilities.

M   **92.** The left hemisphere of the cerebrum is specialized for

    *   a. verbal ability.
       b. spatial relationships.
       c. music interpretation.
       d. control over the left side of the body.

A CLOSER LOOK AT THE CEREBRUM

E   **93.** Which lobe of the cerebrum is a processing center for hearing?

       a. parietal
       b. occipital
    *   c. temporal
       d. frontal
       e. ganglional

E   **94.** Which lobe of the cerebrum process conscious thought and decision making?

       a. parietal
       b. occipital
       c. temporal
    *   d. frontal
       e. ganglional

M   **95.** Emotional states are the responsibility of the

       a. medulla.
       b. corpus callosum.
    *   c. limbic system.
       d. cerebral cortex.
       e. cerebellum.

MEMORY

D   **96.** Studies of memory indicate that

       a. short-term memory is the product of chemical changes in neurons.
       b. long-term memory is limited to a few years' duration.
       c. long-term memories are lost more frequently in amnesia.
    *   d. long-term memory depends on structural or chemical changes in the brain.
       e. short-term memory is limited to several hundred bits of information.

STATES OF CONSCIOUSNESS

M   **97.** An EEG is

    *   a. a graph showing brainwave patterns.
       b. an oscilloscope pattern showing an action potential of a nerve.
       c. an oscilloscope pattern showing a muscle undergoing contraction.
       d. a graphic representation of the function of a heart.
       e. a state of sleep when dreaming occurs.

E   98. The sleep center of the brain is the
        a. pons.
        b. thalamus.
        c. hypothalamus.
    *   d. reticular activating system.
        e. medulla.

M   99. High levels of which chemical in the sleep
        centers of the brain induce drowsiness and
        sleep?
        a. norepinephrine
    *   b. serotonin
        c. adrenalin
        d. enkephalin
        e. cyclic AMP

M   100. States of consciousness are controlled by
        the
    *   a. reticular activating system.
        b. medulla.
        c. cerebellum.
        d. cerebral cortex.
        e. hypothalamus.

EFFECTS OF PSYCHOACTIVE DRUGS

E   101. Drug abuse usually involves those drugs
        which
        a. are psychoactive.
        b. act on the central nervous system.
        c. bind to neuron receptors.
        d. change the chemical messages that
           neurons send and receive.
    *   e. all of these.

M   102. The need for larger and more frequent doses
        of a drug is called
        a. addiction.
        b. habituation.
        c. psychological dependence.
        d. synergism.
    *   e. tolerance.

E   103. In a synergistic drug interaction,
        a. antagonism is increased.
        b. one drug simply enhances the effects of
           the other.
        c. neither drug would work alone.
    *   d. two drugs work additively.
        e. one drug inhibits the other.

E   104. The stimulant in coffee, tea, and soft drinks
        is
        a. nicotine.
        b. sugar.
    *   c. caffeine.
        d. cocaine.
        e. all of these.

M   105. Nicotine mimics _____ in its effects on
        the nervous system.
        a. adrenalin.
    *   b. acetylcholine.
        c. dopamine.
        d. endorphins.
        e. analgesics.

E   106. Amphetamines make person who use them
        feel
        a. very drowsy.
        b. somewhat relaxed.
        c. sedated.
    *   d. more alert.
        e. delirious.

Matching Questions

D **107.** Matching I. Choose the BEST response.

1 _____ effector

2 _____ ganglion

3 _____ integrator

4 _____ stretch receptor

5 _____ myelin sheath

6 _____ neuroglia

7 _____ receptor

8 _____ recovery period

9 _____ response

10 _____ sodium-potassium pump

11 _____ stimulus

12 _____ transmitter substance

    A. cells that nurture and support neurons

    B. a neuron cannot propagate an action potential during this time

    C. ACh

    D. interneuron in brain or spinal cord

    E. establishes basis of resting membrane potential

    F. sheathed muscle cells that contain receptors

    G. input

    H. modified dendrite of a neuron

    I. output

    J. muscle or gland

    K. produced by a specific kind of Schwann cell

    L. cluster of cell bodies from different neurons

Answers:

1. J	2. L	3. D
4. F	5. K	6. A
7. H	8. B	9. I
10. E	11. G	12. C

D **108.** Matching II. Choose the one most appropriate answer for each.

1 _____ autonomic nervous system

2 _____ blood-brain barrier

3 _____ cerebellum

4 _____ ganglion

5 _____ hypothalamus

6 _____ limbic system

7 _____ medulla oblongata

8 _____ midbrain

9 _____ peripheral nervous system

10 _____ reticular activating system

11 _____ transmitter substances

12 _____ white matter

    A. integrates body position, motions, balance

    B. messages from here arouse the brain and maintain wakefulness

    C. feature of the brain capillaries which controls access to neurons

    D. all parts of nerve cells outside the brain and spinal cord

    E. as, for example, acetylcholine

    F. reflex control center for breathing, heart rate, and blood pressure

    G. axons of the central nervous system that are sheathed with fatty myelin

    H. at top of the brain stem bordering the cerebral hemispheres; influences learning and emotional behavior

    I. motor neurons that are divided into sympathetic and parasympathetic divisions

    J. contains the optic lobes; receives and integrates sensory information that is largely sent on to the forebrain for further neural processing

    K. groups of nerve cell bodies encased in connective tissue; form integrative centers

    L. contains centers concerned with body temperature regulation and with salt and water balance

Answers:

1. I	2. C	3. A
4. K	5. L	6. H
7. F	8. J	9. D
10. B	11. E	12. G

Classification Questions

Answer questions 109–113 in reference to the four cell types listed below:

    a. sensory neurons
    b. interneurons
    c. motor neurons
    d. Schwann cells

M   **109.** These nerve cells transmit signals to muscle cells.

M   **110.** This is a type of neuroglial cell.

E   **111.** Myelin is formed by these.

E   **112.** An animal brain is composed mostly of this cell type.

E   **113.** This cell type picks up environmental signals.

Answers: 109. c   110. d   111. d
                 112. b   113. a

Answer questions 114–118 in reference to the autonomic nervous system associated with the five regions of the human vertebral column listed below:

    a. cervical
    b. thoracic
    c. lumbar
    d. sacral
    e. coccygeal

D   **114.** Sympathetic nerves from this region innervate the bladder, uterus, and genitals.

D   **115.** Sympathetic nerves from this region innervate the kidney.

D   **116.** Parasympathetic nerves from this region innervate the rectum.

D   **117.** Sympathetic nerves from this region innervate the heart.

D   **118.** Sympathetic nerves from this region pass through the celiac ganglion.

Answers: 114. c   115. b   116. d
                 117. b   118. b

Answer questions 119–123 in reference to the five regions of the vertebrate brain listed below:

    a. cerebrum
    b. hypothalamus
    c. pons
    d. cerebellum
    e. medulla oblongata

D   **119.** This region of the brain contains the reflex centers involved in respiration.

D   **120.** This region of the brain controls neural-endocrine activities such as temperature control.

D   **121.** This region of the brain controls carbohydrate metabolism.

M   **122.** This part of the brain controls the complex coordination of motor activity and limb movement.

D   **123.** This region of the brain contains parasympathetic nerves that innervate the heart and lungs.

Answers: 119. e   120. b   121. b
                 122. d   123. e

Selecting the Exception

M   **124.** Four of the five answers listed below are actively involved in nerve impulse transmission. Select the exception.
    * a. neuroglia
      b. neuron
      c. ganglia
      d. nerves
      e. tracts

M   **125.** Four of the five answers listed below are true of a neuron at resting potential. Select the exception.
    * a. interior of neuron positive
      b. interior of neuron has negative charge
      c. more sodium ion outside neuron
      d. more potassium ions inside neuron
      e. membrane of neuron is polarized

M   **126.** Four of the five answers listed below are used in descriptions of neuron membranes. Select the exception.
      a. gate
      b. pump
      c. wave of depolarization
      d. channel
    * e. synaptic cleft

M 127. Four of the five answers listed below are used in descriptions of the nerve sheath. Select the exception.
 a. Schwann cell
* b. threshold
 c. myelin sheath
 d. saltatory conduction
 e. node

M 128. Four of the five answers listed below are participants in a common function. Select the exception.
 a. sensory neuron
* b. medulla
 c. interneuron
 d. effector
 e. receptor

M 129. Four of the five answers listed below are parts of the central nervous system. Select the exception.
 a. spinal cord
 b. medulla
* c. ganglia
 d. cerebellum
 e. cerebrum

M 130. Four of the five answers listed below are parts of the same nerve grouping. Select the exception.
* a. cranial nerve
 b. thoracic nerve
 c. lumbar nerve
 d. cervical nerve
 e. sacral nerve

E 131. Four of the five answers listed below are innervated by the autonomic nervous system. Select the exception.
* a. skeletal muscles
 b. smooth muscles
 c. heart
 d. endocrine glands
 e. exocrine glands

D 132. Four of the five answers listed below are actions mediated by the sympathetic nervous system. Select the exception.
 a. pulse increase
* b. blood glucose levels drop
 c. metabolism increases
 d. digestion slows down
 e. dilates pupils of the eye

D 133. Four of the five answers listed below are located within the spinal cord. Select the exception.
* a. dorsal root ganglion
 b. interneurons between sensory input motor outputs
 c. major ascending and descending nerve tracts
 d. interneurons connecting with other neurons
 e. direct reflex connections between sensory and motor neurons

M 134. Four of the five answers listed below are parts of the forebrain. Select the exception.
 a. limbic system
 b. thalamus
 c. olfactory lobes
* d. cerebellum
 e. cerebrum

D 135. Four of the five answers listed below are classified as white matter. Select the exception.
 a. myelin sheath
 b. descending tracts
 c. corpus callosum
* d. cerebrum
 e. ascending tract

M 136. Four of the five answers below are not legitimate prescription drugs. Select the exception.
* a. barbiturates
 b. PCP
 c. amphetamine
 d. cocaine
 e. alcohol

CHAPTER 12
SENSORY RECEPTION

Multiple-Choice Questions

SENSORY RECEPTORS AND PATHWAYS—AN OVERVIEW

M   **1.** A sensory system includes
   a. a sensory receptor.
   b. nerve pathways from the receptor to the brain.
   c. brain regions where the sensory information is processed.
   d. a sensory receptor and nerve pathways from the receptor to the brain, only.
 * e. a sensory receptor, nerve pathways from the receptor to the brain, and brain regions where the sensory information is processed.

D   **2.** The difference in "sensation" and "perception" when referring to a stimulus lies in
   a. the type of receptor stimulated.
 * b. understanding the significance of the stimulus.
   c. the number of receptors that depolarize.
   d. feeling exactly what is happening at the site.
   e. responding to the stimulus.

M   **3.** Which of the following BEST defines a sensory receptor?
   a. an organ, such as the human eye
 * b. endings of sensory neurons
   c. highly branched endings of axons
   d. any form of energy
   e. an action potential

M   **4.** Which of the following BEST defines a stimulus?
   a. an action potential in a neuron
   b. depolarization of a nerve cell
 * c. energy that elicits a response
   d. the recording of an event in the brain

E   **5.** Which sense utilizes mechanical energy?
   a. sense of touch
   b. muscle sense
   c. sense of hearing
   d. sense of balance
 * e. all of these

E   **6.** Olfactory centers are responsive to
   a. touch.
 * b. smell.
   c. taste.
   d. sound.
   e. sight.

M   **7.** Nociceptors are involved with the detection of
   a. heat energy.
   b. changes in body position.
   c. chemicals in the air or water.
 * d. pain.
   e. visible light.

E   **8.** The carotid body detects carbon dioxide in the blood and is therefore a
 * a. chemoreceptor.
   b. photoreceptor.
   c. thermoreceptor.
   d. mechanoreceptor.
   e. movement receiver.

M   **9.** The major function of a receptor is to
   a. control the autonomic functions of the body.
   b. stabilize the internal environment to achieve homeostasis.
   c. produce responses to the various stimuli the body receives.
 * d. give organisms awareness and sensitivity to their environment.
   e. interpret the sensations it receives from the environment.

D   **10.** The extent to which a sense can detect a particular sensation is mainly due to the
   a. sensitivity of its receptors.
   b. speed of nerve transmission.
 * c. brain area devoted to interpretation.
   d. transmitter substances in the synapses.
   e. amount of stimulus.

D 11. Which of the following statements is false?
   a. Sensory receptors are able to detect specific stimuli only.
   b. Sensation is an awareness of a change in external or internal conditions.
 * c. A stimulus will generate impulses that are different depending on where the signals are sent in the brain.
   d. Information about a stimulus is coded in the quantity and frequency of action potentials sent to the brain.
   e. Specific regions of the brain translate the information about the signal and convert it into a sensation.

M 12. Which of the following is NOT characteristic of sensation?
   a. Some sense cells can be activated only by chemicals.
   b. Nerve impulses for each of the five senses are essentially the same regardless of the sense considered.
   c. Sensory nerve impulses follow the all-or-none law.
 * d. All organisms perceive their environments by the same sense organs.

M 13. Differences in intensity of a stimulus
   a. do not affect the impulse transmitted.
   b. are indicated by the number of nerves activated.
   c. control the part of the brain that receives the stimulus.
   d. are encoded in the frequency of action potentials on a single axon.
 * e. are indicated by the number of nerves activated and are encoded in the frequency of action potentials on a single axon.

D 14. Which of the following statements is false?
   a. It is assumed that lower animals are incapable of perception.
 * b. Humans, as the most complex organisms, have all of the sensory organs found in lower forms.
   c. Mechanoreceptors are responsible for the senses of equilibrium and hearing.
   d. Infrared radiation is detected by thermoreceptors.
   e. A lobster can react to the boiling water in which it is placed to cook.

M 15. Which of the following statements is true?
   a. All action potentials for the same nerve are alike.
   b. Receptors respond by graded potentials.
   c. Different nerves convey different information by going to different areas of the brain.
   d. Differences in intensities may be due to the number of receptors and nerves involved in response.
 * e. all of these

M 16. A loud sound can be distinguished from a soft whisper because
   a. more neurons depolarize.
   b. each receptor depolarizes more fully.
   c. the frequency of depolarizing neurons increases.
 * d. more neurons depolarize and the frequency of depolarizing neurons increases.
   e. more neurons depolarize, each receptor depolarizes more fully, and the frequency of depolarizing neurons increases.

M 17. Which of the following statements is false?
   a. Neurons in the brain can interpret incoming action potentials only in certain ways.
   b. Optic nerve impulses can only be interpreted as light.
 * c. Sensory neurons do not follow the all-or-none law.
   d. The stronger the stimulus, the more action potentials are generated.
   e. A strong stimulus activates many adjacent receptors, thereby increasing the number of action potentials and the level of awareness for the stimulus.

D 18. Which of the following is NOT an example of nerve adaptation?
   a. a loss of the sense of pressure of clothes against the skin
   b. adjustment to repeated jumping into a body of cold water
 * c. response to lack of oxygen at high altitudes
   d. eventual loss of awareness of a constant noise
   e. the loss of the ability to smell a person's perfume or cologne after being in that person's presence for some time

SOMATIC SENSATIONS

E   **19.** The Pacinian corpuscle is used in detecting
   - a. sound.
   - * b. pressure.
   - c. chemicals.
   - d. sight.
   - e. chemical differences.

M   **20.** Which of the following statements is false?
   - * a. All animals placed in the same environment will have the same awareness of it.
   - b. Humans and insects see flowers differently.
   - c. The carotid bodies monitor the concentration of carbon dioxide in the blood.
   - d. Olfactory receptors detect odors.
   - e. Pacinian corpuscles are examples of mechanoreceptors.

M   **21.** The somatic senses include all but which one of the following sensations?
   - * a. balance
   - b. pain near the body surface
   - c. temperature
   - d. touch
   - e. pressure

M   **22.** Somatic sensations include all but which one of the following?
   - a. heat and cold
   - b. pressure and touch
   - c. pain
   - d. limb movements and the position of the body in space
   - * e. sound

M   **23.** According to the classification given by your authors, somatic senses
   - * a. are distributed in several locations over the body.
   - b. reside in certain receptor organs located at a few specific locations.
   - c. are exemplified by ears and eyes.
   - d. include sight and sound.

M   **24.** Mechanoreceptors are located in
   - a. internal organs.
   - b. skin.
   - c. joints.
   - d. tendons.
   - * e. all of these

D   **25.** The feeling of pressure on the skin is the result of
   - a. bending of mechanoreceptors.
   - b. stimulation only when the stimulus is first applied.
   - c. constant stimulation.
   - * d. bending of mechanoreceptors and constant stimulation.
   - e. bending of mechanoreceptors and stimulation only when the stimulus is first applied.

E   **26.** Which of the following statements is true concerning thermoreception?
   - a. When the temperature at the body surface remains stable, the thermoreceptors become inactive.
   - b. When the body temperature increases, receptors decrease their rate of firing.
   - c. Heat and cold are received by the same receptors..
   - * d. Increased depolarization of thermoreceptors is the result of increased temperatures.
   - e. Pacinian corpuscles are important thermoreception devices.

M   **27.** Pain is
   - a. one of the special senses.
   - b. the perception of injury.
   - c. dependent on interpretation by the brain.
   - * d. the perception of injury and dependent on interpretation by the brain.
   - e. one of the special senses, the perception of injury, and dependent on interpretation by the brain.

E   **28.** The pain produced in an internal organ may be perceived as occurring somewhere else. This is called
   - a. mixed nerve messages.
   - * b. referred pain.
   - c. phantom pain.
   - d. psychosomatic pain.
   - e. hypochondria.

E   **29.** A stretch receptor is classified as a
   - a. chemoreceptor.
   - * b. mechanoreceptor.
   - c. photoreceptor.
   - d. thermoreceptor.
   - e. all of these

D 30. The knee-jerk reflex used by physicians to check nerve response is based on
   a. muscle spindles.
   b. stretch receptors.
   c. spinal cord synapses.
   d. muscle spindles and stretch receptors, only.
 * e. muscle spindles, stretch receptors, and spinal cord synapses.

SENSES OF TASTE AND SMELL

E 31. Receptors in the human nose are
 * a. chemoreceptors.
   b. mechanoreceptors.
   c. photoreceptors.
   d. nociceptors.
   e. none of these

E 32. Sense receptors for "taste" are located
   a. on the tongue.
   b. on the roof of the mouth.
   c. in the throat
   d. on the palate.
 * e. all of these

M 33. Which of the following is NOT a basic taste category identified in humans?
   a. bitter
   b. salty
 * c. spicy
   d. sour
   e. sweet

E 34. Interpretation of smell is accomplished
   a. by the nasal epithelium.
   b. in the olfactory receptors.
   c. by centers in the brainstem.
 * d. olfactory bulbs in the brain.
   e. independently of olfactory receptors.

M 35. Functionally, the two most closely associated senses are
   a. sight and sound.
   b. touch and sight.
 * c. taste and smell.
   d. temperature and pain.
   e. touch and balance.

SENSE OF HEARING

E 36. The sense based on air vibrations is
   a. taste.
   b. smell.
   c. touch.
   d. sight.
 * e. hearing.

E 37. The organ of Corti is a
   a. chemoreceptor.
 * b. mechanoreceptor.
   c. photoreceptor.
   d. nociceptor.
   e. all of these

M 38. The principal place in the human ear where sound waves are amplified by means of the vibrations of tiny bones is the
   a. pinna.
   b. ear canal.
 * c. middle ear.
   d. organ of Corti.
   e. all of these

M 39. The place where vibrations are translated into patterns of nerve impulses is the
   a. pinna.
   b. ear canal.
   c. middle ear.
 * d. organ of Corti.
   e. tympanum

E 40. The organ of Corti is located in the
   a. thoracic cavity.
 * b. inner ear.
   c. abdominal cavity.
   d. brain stem.
   e. semicircular canals.

D 41. In hearing, the last place that pressure or sound waves pass through is the
   a. bones of the middle ear.
   b. tympanic membrane.
   c. oval window.
 * d. round window.
   e. tectorial membrane.

E 42. How many coiled and fluid-filled ducts are found in each cochlea?
   a. 1
   b. 2
 * c. 3
   d. 4
   e. 5 or more

D   43. The sense of hearing in vertebrates is dependent on
    a. fluid displacement.
    b. hair bending.
    c. bone movement.
    d. membrane vibration.
  * e. all of these

D   44. The sense in which amplitude and frequency can be detected with some accuracy is
    a. sight.
  * b. hearing.
    c. balance.
    d. taste.
    e. smell.

M   45. The equalization of pressures between the ear and throat is made possible by the
    a. pharynx.
  * b. Eustachian tube.
    c. cupula.
    d. semicircular canals.
    e. round window.

E   46. Movable bones are features of the sense organs associated with
    a. sight.
  * b. hearing.
    c. taste.
    d. smell.
    e. touch.

M   47. Movable bones are found in the
    a. cochlea.
    b. external ear.
  * c. middle ear.
    d. inner ear.
    e. organ of Corti.

M   48. The organ of Corti
    a. functions in the awareness of motion and the sense of equilibrium.
    b. controls the sense of depth perception.
  * c. converts sound vibrations into impulses that enable hearing.
    d. secretes cerebrospinal fluid.
    e. detects light energy.

SENSE OF BALANCE

M   49. Hair cells are important in the sense of
    a. equilibrium.
    b. hearing.
    c. taste.
    d. smell.
  * e. both equilibrium and hearing.

E   50. One's equilibrium is sensed by a
    a. chemoreceptor.
  * b. mechanoreceptor.
    c. photoreceptor.
    d. thermoreceptor.
    e. none of these

E   51. The sense of equilibrium or balance can detect
    a. motion.
    b. acceleration.
    c. gravity.
    d. position.
  * e. all of these

M   52. The semicircular canals are
    a. empty.
    b. filled with gas.
  * c. filled with a liquid.
    d. filled with bones or stones.
    e. filled with sand grains.

E   53. An otolith is one of the functional parts of the
    a. eye.
    b. Pacinian corpuscle.
  * c. vestibular apparatus.
    d. taste bud.
    e. pits, or heat-sensing devices, of snakes.

E   54. How many semicircular canals are in each organ of balance?
    a. 2
  * b. 3
    c. 4
    d. 5
    e. more than 6

D   55. Motion sickness is the result of
    a. overstimulation of the hair cells in the vestibular apparatus.
    b. visual input, especially when it is monotonous.
    c. fear and anxiety.
    d. overstimulation of the hair cells in the vestibular apparatus and visual input, especially when it is monotonous.
  * e. overstimulation of the hair cells in the vestibular apparatus and visual input, especially when it is monotonous, plus fear and anxiety.

VISION: AN OVERVIEW

E   **56.** Eyes are
      a. chemoreceptors.
      b. mechanoreceptors.
    * c. photoreceptors.
      d. nociceptors.
      e. none of these

E   **57.** The layer of the eye where photoreceptors
       are located is the
      a. lens.
      b. cornea.
      c. pupil.
      d. iris.
    * e. retina.

E   **58.** The adjustable ring of contractile and
       connective tissues that controls the amount
       of light entering the eye is the
      a. lens.
      b. cornea.
      c. pupil.
    * d. iris.
      e. retina.

E   **59.** The white protective fibrous tissue of the
       eye, often called the white of the eye, is the
      a. lens.
    * b. sclera.
      c. pupil.
      d. iris.
      e. retina.

E   **60.** The dark middle layer of the eye that
       prevents the scattering of light is the
      a. fovea.
      b. retina.
      c. sclera.
    * d. choroid.
      e. cornea.

M   **61.** Accommodation involves the ability to
      a. change the sensitivity of the rods and
         cones by means of neurotransmitters.
      b. change the curvature of the cornea.
    * c. change the width of the lens by
         contracting or relaxing certain muscles.
      d. adapt to large changes in light
         intensity.
      e. all of these

M   **62.** The ciliary muscle
      a. controls the eardrum.
    * b. controls the shape of the lens to allow
         focusing.
      c. holds the bones of the middle ear in
         place.
      d. enables the eyeball to move so that a
         person may see an object without
         moving the head.
      e. is responsible for the size of the pupil
         in different light intensities.

M   **63.** If the ciliary muscle of the eye is damaged,
       then
      a. color vision will be lost.
      b. the amount of light entering the eye
         cannot be regulated.
    * c. proper focusing will be impossible.
      d. only peripheral vision is available.
      e. "night blindness" will become more
         evident.

E   **64.** The outer transparent protective cover over
       the front of the eyeball is the
      a. fovea.
      b. retina.
      c. sclera.
      d. choroid.
    * e. cornea.

E   **65.** The part of the eye that may be colored
       (e.g., brown, blue, green, or gray) is the
      a. retina.
      b. sclera.
      c. choroid.
      d. cornea.
    * e. iris.

FROM NEURON SIGNALING TO VISUAL PERCEPTION

E   **66.** Rods and cones are located in the
      a. lens.
      b. cornea.
      c. pupil.
      d. iris.
    * e. retina.

E   **67.** Cones are
      a. sensitive to red light.
      b. sensitive to green light.
      c. sensitive to blue light.
      d. relatively insensitive to dim light.
    * e. all of these

M   68. Where are bipolar, amacrine, and ganglion cells located?
     a. sclera
     b. thalamus
     c. organ of Corti
*   d. retina
     e. all of these

E   69. Rhodopsin is a molecule made of
     a. protein only.
     b. carbohydrate plus protein.
     c. vitamin and protein derivative.
     d. steroid.
*   e. protein and vitamin derivative.

E   70. The highest concentration of cones is in the
*   a. fovea.
     b. blind spot.
     c. sclera.
     d. iris
     e. choroid.

D   71. The fovea
     a. is the blind spot produced by the optic nerve entering the eye.
*   b. is the region of the retina filled with cones that allows the most acute vision.
     c. is the region of the retina that has the greatest concentration of rods that enable sight under extremely dim conditions.
     d. focuses light on the retina.
     e. is the anterior fluid-filled chamber of the eye.

M   72. In the human eye, what provides the greatest visual acuity (the precise discrimination between adjacent points in space)?
     a. photoreceptors in the sclera
*   b. photoreceptors in the fovea
     c. protein filaments in the lens
     d. photoreceptors in the optic nerve
     e. none of these

Matching Questions

D   73. Matching. Choose the one appropriate answer for each.
     1. ____ amplitude
     2. ____ cochlea
     3. ____ eardrum
     4. ____ iris
     5. ____ oval window
     6. ____ organ of Corti
     7. ____ malleus
     8. ____ pitch
     9. ____ retina
     10. ____ round window
     11. ____ semicircular canals

     A. dissipates excess vibrational energy to the middle ear
     B. contains the organ of Corti
     C. separates the outer and middle ears
     D. membrane-covered gateway to inner ear
     E. consists of tissue containing rods and cones
     F. bone in the middle ear
     G. peak height and valley depth of sound waves are its basis
     H. maintain balance and position; detect acceleration
     I. depends on how many wave changes per second occur
     J. contains hair cells used in sound reception
     K. regulates size of pupil and amount of incoming light

Answers: 1. G   2. B   3. C
            4. K   5. D   6. J
            7. F   8. I   9. E
            10. A   11. H

Classification Questions

Answer questions 74–78 in reference to the five kinds of energy listed below:

      a. chemical
      b. mechanical
      c. thermal
      d. light
      e. wavelike form of mechanical energy

E   **74.** Receptors on the tongue detect variation in this kind of energy.

M   **75.** Olfactory receptors detect this kind of energy.

D   **76.** Ears monitor this kind of energy.

E   **77.** Human eyes respond to this kind of energy.

M   **78.** The pain you feel from a pebble in your shoe is a result of detecting this kind of energy.

Answers: 74. a   75. a   76. e
               77. d   78. b

Answer questions 79–83 in reference to the five eye structures listed below:

      a. cornea
      b. lens
      c. retina
      d. ciliary body
      e. vitreous body

D   **79.** This structure changes the shape of the lens to focus light.

M   **80.** This structure is composed of photoreceptor cells.

E   **81.** This structure primarily acts to focus light waves.

M   **82.** This structure is composed of rod- and cone-shaped cells in mammals and birds.

M   **83.** This structure acts to maintain the shape of the eye and to transmit light to other structures.

Answers: 79. d   80. c   81. b
               82. c   83. e

Selecting the Exception

D   **84.** Four of the five answers listed below are related by a similar sense receptor. Select the exception.
        a. touch or pressure
  *   b. olfaction
        c. balance (equilibrium)
        d. hearing
        e. muscle sense

D   **85.** Four of the five answers listed below are related by a similar sensation. Select the exception.
        a. incus
        b. organ of Corti
        c. tectorial membrane
        d. oval window
  *   e. Pacinian corpuscle

M   **86.** Four of the five answers listed below are somatic senses. Select the exception.
  *   a. light
        b. pressure
        c. touch
        d. temperature
        e. pain

D   **87.** Four of the five answers listed below are related by a similar receptor. Select the exception.
        a. otolith
        b. vestibular apparatus
        c. semicircular canal
  *   d. cochlea
        e. saccule

M   **88.** Four of the five answers listed below are parts of the inner ear. Select the exception.
  *   a. eardrum
        b. oval window
        c. scala tympani
        d. basilar membrane
        e. cochlea

D   **89.** Three of the four answers listed below are functionally connected to each other. Select the exception.
        a. hammer
  *   b. pinna
        c. stirrup
        d. anvil

M   **90.** Four of the five answers listed below are
           parts of the same sense organ. Select the
           exception.
           a. choroid
           b. retina
           c. vitreous humor
       *   d. ampulla
           e. sclera

M   **91.** Four of the five answers listed below are
           parts of the same sense organ. Select the
           exception.
       *   a. cochlea
           b. cornea
           c. sclera
           d. choroid
           e. fovea

D   **92.** Three of the four answers listed below are
           colors for which the cone cells have
           pigments. Select the exception.
           a. red
       *   b. yellow
           c. blue
           d. green

M   **93.** Four of the five answers listed below are
           receptors. Select the exception.
       *   a. muscle
           b. Pacinian corpuscle
           c. hair cells
           d. cones
           e. taste buds

CHAPTER 13

THE ENDOCRINE SYSTEM

Multiple-Choice Questions

THE ENDOCRINE SYSTEM

M **1.** When Bayliss and Starling conducted experiments to determine what caused secretion of pancreatic juice, they
   a. cut the blood supply to the upper digestive tract.
   b. added acid to the small intestine and got secretions of pancreatic juices.
   c. cut the nerve supply to the upper digestive tract.
   d. got a response from the pancreas using extracts of cells lining the intestinal tract.
   * e. All except "cut the blood supply to the upper digestive tract" are true

E **2.** The first hormone to be discovered was
   a. insulin.
   b. gastrin.
   * c. secretin.
   d. thyroxine.
   e. estrogen.

E **3.** The word *hormone* comes from the Greek word meaning
   a. target.
   b. response.
   c. secretion.
   * d. set in motion.
   e. internal gland.

E **4.** Most hormones are distributed throughout the body by the
   a. exocrine system.
   b. lymphatic system.
   c. nervous system.
   * d. blood system.
   e. integumentary system.

M **5.** The primary purpose of the endocrine system is to
   a. provide a mechanism for rapid response to changes in the body.
   * b. maintain a relatively constant internal environment.
   c. ensure proper growth and development.
   d. allow for a mechanism to control gene action.
   e. all of these

M **6.** Target cells
   a. are found only in specific endocrine glands.
   b. are equipped with specific receptor molecules.
   c. are muscle cells.
   d. may occur in any part of the body.
   * e. are equipped with specific receptor molecules and may occur in any part of the body.

M **7.** Which of the following statements is true?
   * a. Although hormones are carried to all parts of the body, they produce effects only in cells with proper receptors.
   b. Hormones are limited to steroid compounds.
   c. Hormones are secreted by specialized exocrine glands.
   d. Most hormones are controlled by positive feedback mechanisms involving the pituitary gland.

M **8.** The important feature of all cells that react to a specific hormone is the
   a. type of blood supply they receive.
   b. proximity of the endocrine gland.
   * c. presence of an appropriate receptor molecule.
   d. characteristics of their plasma membranes.
   e. presence of specific genes responsive to the hormone.

M **9.** In contrast to hormones, neurotransmitters
   a. are released from neurons.
   * b. not carried by the bloodstream.
   c. derivatives of peptides.
   d. bind only to specific receptors.

SIGNALING MECHANISMS

M **10.** In the testicular feminization syndrome
   a. no testosterone is produced.
   b. chemicals circulating in the blood deactivate the male hormone.
   * c. the cellular receptor for testosterone in the target cells is defective.
   d. the male with this defect is normal in all respects except that he is sterile.

D (11.) The reason that some individual hormones have so many different effects is that
   a. they influence gene transcription.
   b. they trigger a second messenger system that produces a cascade of effects.
 * c. there are a great many different cells in different tissues that have specific receptors for the hormone.
   d. the hormone is carried throughout the body and only a small amount is needed to produce its effect.
   e. all of these

M 12. Steroid hormones do not require a membrane receptor because they
   a. are small enough to pass directly through the membrane.
 * b. are lipid-soluble in the bilayer.
   c. pass through special channels.
   d. are water-soluble.
   e. dissolve in the cholesterol of the membranes.

M 13. Which is the predominant second messenger involved in regulating glucose metabolism?
   a. insulin
   b. glucagon
   c. adenyl cyclase
 * d. cyclic AMP
   e. all of these

M 14. Second messengers are molecules of
   a. steroid compounds.
 * b. cyclic AMP.
   c. ADP.
   d. prostaglandin.
   e. intermedin.

M (15) Water-soluble hormones
   a. have to be transported by specific carriers in the blood.
   b. have no trouble entering the target cells.
   c. find and react with the surface receptor molecules.
   d. sometimes elicit the production of a second messenger.
 * e. all except "have no trouble entering the target cells."

THE HYPOTHALAMUS AND PITUITARY GLAND

E 16. Which gland is often called the master gland?
   a. pineal
 * b. pituitary
   c. thyroid
   d. adrenal
   e. pancreas

E 17. The pituitary gland is controlled by the
   a. pons.
   b. corpus callosum.
   c. medulla.
   d. thalamus.
 * e. hypothalamus.

M 18. Which is an example of an organ that is nervous in origin, structure, and function but secretes substances into the bloodstream?
   a. anterior pituitary
 * b. posterior pituitary
   c. pancreas
   d. adrenal cortex
   e. testis

D 19. The hypothalamus and pituitary link the activities of the endocrine system and nervous system by
 * a. neurohormones being secreted in response to the summation of neural messages that enter the hypothalamus.
   b. shifts in hormonal concentrations being detected by the anterior pituitary.
   c. pheromones being secreted as a response to photoperiodic stimuli.
   d. the nervous tissue of the anterior lobe of the pituitary sending stimuli to the glandular tissue of the posterior pituitary to produce hormones that will be secreted by the hypothalamus.
   e. all of these

M 20. Which statement is true?
   a. The anterior pituitary gland is essentially nervous tissue.
   b. The anterior pituitary gland secretes only two hormones.
   c. The posterior pituitary gland is the master gland.
 * d. The posterior pituitary gland only stores hormones produced by the hypothalamus.
   e. all of these

E   **21.** If you were cast up on a desert island with no fresh water to drink, the level of which of the following would rise in your bloodstream in an effort to conserve water?
     a. erythropoietin
     b. oxytocin
     c. insulin
   *   d. antidiuretic hormone
     e. glucose

M   **22.** The antidiuretic hormone
     a. controls water balance.
     b. controls the concentration of urea in the urine.
     c. influences blood pressure.
     d. changes the permeability of the urine-conducting tubules so that the interstitial fluid increases.
   *   e. all of these

E   **23.** Oxytocin affects the
   *   a. uterine wall.
     b. voluntary muscles throughout the body.
     c. nervous tissue.
     d. target cells in the brain.
     e. target cells in the digestive tract.

M   **24.** The control over milk production, water balance, and labor in childbirth is mediated by the _____ gland.
     a. pineal
     b. anterior pituitary
   *   c. posterior pituitary
     d. parathyroid
     e. thyroid

E   **25.** The immediate stimulus for the release of milk from the female breast is the
     a. culmination of the maternal instinct.
     b. excessive accumulation of milk.
   *   c. mechanical stimulation of the breast by sucking.
     d. time of day.
     e. interaction of hormones.

D   **26.** Antidiuretic hormone and oxytocin are products of
     a. endocrine glands.
   *   b. neurosecretory cells.
     c. blood capillaries.
     d. the anterior pituitary.
     e. kidney and uterine wall cells, respectively.

M   **27.** A drop in blood volume would trigger the body to secrete
     a. parathyroid hormones.
     b. somatotropin.
   *   c. antidiuretic hormone.
     d. insulin.
     e. glucocorticoids.

E   **28.** The anterior pituitary secretions produce their effects in the
     a. gonads.
     b. thyroid glands.
     c. adrenal glands.
     d. mammary glands.
   *   e. all of these

E   **29.** ACTH
     a. is secreted by the posterior pituitary.
     b. has target cells in the autonomic nervous system.
   *   c. has target cells in the adrenal cortex.
     d. has target cells in the adrenal medulla.
     e. initiates the autoimmune response.

M   **30.** The pituitary hormone associated most directly with metabolic rate and with growth and development is
     a. ACTH.
   *   b. TSH.
     c. FSH.
     d. LH.
     e. ADH.

M   **31.** The luteinizing hormone
   *   a. stimulates ovulation.
     b. has no function in males.
     c. is produced by the corpus luteum.
     d. stimulates milk production.
     e. promotes sperm formation.

M   **32.** The most general of the pituitary hormones, in that it may affect almost any cell in the body, is
     a. the adrenocorticotropic hormone.
     b. the thyroid-stimulating hormone.
     c. gonadotropin.
   *   d. somatotropin.
     e. prolactin.

M   33. Prolactin
    *   a. stimulates the mammary glands to
            produce milk.
        b. causes the development of breasts and
            other secondary sexual characteristics in
            the male.
        c. acts in concert with FSH to produce
            milk.
        d. has secondary effects on reducing the
            size of the uterus after birth.

E   34. The growth hormone is
        a. prolactin.
        b. adrenalin.
        c. thyroxine.
        d. ACTH.
    *   e. somatotropin.

D   35. Which of the following hormones is
        different from the others based upon the
        range of its target cells?
        a. corticotropin
        b. luteinizing hormone
    *   c. somatotropin
        d. thyrotropin
        e. follicle-stimulating hormone

D   36. The secretion of each of the hormones from
        the anterior pituitary requires
        a. stimulation from the posterior
            pituitary.
        b. that they first be secreted from the
            neurons of the hypothalamus.
        c. two capillary beds.
        d. the action of minute amounts of
            releasing hormones.
    *   e. two capillary beds and the action of
            minute amounts of releasing hormones.

EXAMPLES OF ABNORMAL PITUITARY OUTPUT

E   37. Dwarfism may be due to insufficient
        production of
        a. mineralocorticoid.
        b. glucocorticoid.
        c. calcitonin.
    *   d. somatotropin.
        e. the parathyroid hormone.

E   38. Acromegaly is the result of excessive
        secretion of which of the following by
        adults?
        a. mineralocorticoid
        b. glucocorticoid
        c. thyroxine
        d. testosterone
    *   e. somatotropin

SOURCES AND EFFECTS OF OTHER HORMONES

M   39. Which hormone prepares and maintains the
        uterine lining for pregnancy?
        a. estrogen
        b. progesterone
        c. follicle-stimulating hormone
        d. luteinizing hormone
    *   e. both estrogen and progesterone

E   40. Angiotensin is produced in the
        a. bloodstream.
        b. adrenal cortex.
        c. adrenal medulla.
    *   d. kidneys.
        e. heart.

D   41. You have just moved from Norfolk,
        Virginia (sea level), to Taos, New Mexico
        (high in the mountains), and you find
        yourself out of breath climbing a small hill.
        Three months later, climbing the same hill,
        you have no difficulty. In the interim you
        have not altered your level of activity or
        diet. Which hormone has been at work?
        a. angiotensin
    *   b. erythropoietin
        c. aldosterone
        d. estrogen
        e. none of these

E   42. The gonads are another name for the
        a. parathyroid and thyroid.
    *   b. ovary and testis.
        c. adrenal cortex and medulla.
        d. anterior and posterior pituitary.
        e. none of these

E   43. Which gland secretes sex hormones?
    *   a. testis
        b. adrenal medulla
        c. thyroid
        d. kidney
        e. pancreas

D   44. Which of the following is NOT produced by the gonads?
   a. testosterone
   b. progesterone
* c. follicle-stimulating hormone
   d. androgens
   e. All are produced by the gonads.

FEEDBACK CONTROL OF HORMONE ACTIVITY—SOME EXAMPLES

E   45. Which of the following is a hormone secreted by the adrenal cortex that regulates mineral balance?
* a. aldosterone
   b. epinephrine
   c. antidiuretic hormone
   d. angiotensin
   e. insulin

M   46. Glucocorticoids
   a. are secreted by the adrenal cortex.
   b. influence carbohydrate, fat, and protein metabolism.
   c. function during infection and injury as part of the defense response.
   d. are exemplified by cortisol.
* e. all of these

D   47. The first hormone produced by the body in response to an attack of hypoglycemia is
   a. adrenocorticotropin.
   b. glucocorticoid.
* c. corticotropin releasing hormone.
   d. insulin.
   e. thyroxine.

D   48. A friend tells you that her husband has been feeling guilty and stressed for the past month. During the same time interval he has felt fatigued most of the time and many foods now seem to upset his stomach. Doctors have already checked for ulcers, cancer, blood pressure changes, and other blood irregularities, but these apparently are normal. You are an endocrinologist, so you suggest that he be tested for the most likely endocrine malfunction, which would be
   a. androgen/estrogen levels in the bloodstream.
* b. glucocorticoid levels in the bloodstream.
   c. calcitonin level in the bloodstream.
   d. melatonin level in the bloodstream.
   e. none of these

D   49. Blood glucose levels are regulated by
   a. insulin.
   b. glucagon.
   c. cortisol.
   d. insulin and glucagon, only.
* e. insulin, glucagon, and cortisol.

M   50. The hormone whose levels remain high when the body is suffering from inflammation and stress is
* a. cortisol.
   b. somatotropin.
   c. thymosin.
   d. prolactin.
   e. parathyroid hormone.

E   51. The adrenal medulla produces
   a. mineralocorticoids.
* b. epinephrine.
   c. cortisol.
   d. testosterone.
   e. glucocorticoids.

D   52. The only endocrine gland whose secretory function is under direct control by sympathetic nerves is the
   a. pancreas.
   b. thyroid.
* c. adrenal medulla.
   d. thymus.
   e. testis.

M   53. The "fight-or-flight" response is enhanced by secretions
   a. from the adrenal cortex.
* b. known as epinephrine and norepinephrine.
   c. stored in the pancreas.
   d. from the adrenal cortex, as well as epinephrine and norepinephrine.
   e. from the adrenal cortex, as well as epinephrine and norepinephrine, and are stored in the pancreas.

E   54. A goiter is an enlarged form of which gland?
   a. adrenal
   b. pancreas
* c. thyroid
   d. parathyroid
   e. thymus

D   55. A goiter is caused by a deficiency in
   a. thyroxine.
   b. triiodothyronine.
   c. calcium.
* d. iodine.

E 56. Synthetic thyroxine taken orally in the form
   of pills is the recommended treatment for
   *   a. hypothyroidism.
       b. goiter.
       c. hyperthyroidism.
       d. Grave's disease.
       e. acromegaly.

HORMONE RESPONSES TO LOCAL CHEMICAL CHANGES

M 57. Calcitonin acts in opposition to
   *   a. the parathyroid hormone.
       b. thyroxine.
       c. glucagon.
       d. the adrenal medulla.
       e. all of these

M 58. If you eliminated all sources of calcium
   (dairy products, some vegetables) from your
   diet, the level of which of the following
   would rise in an attempt to supply calcium
   stored in your body to the tissues that need
   it?
       a. aldosterone
       b. calcitonin
       c. mineralocorticoids
   *   d. parathyroid hormone
       e. all of these

M 59. Which of the following does NOT affect
   blood sugar levels?
       a. glucagon
       b. epinephrine
   *   c. parathyroid hormones
       d. glucocorticoids
       e. insulin

E 60. The normal individual has how many
   parathyroid glands?
       a. 2
       b. 3
   *   c. 4
       d. 5
       e. 6

M 61. Which of the following glands produces
   only one type of hormone?
       a. adrenal medulla
       b. adrenal cortex
       c. pancreas
   *   d. parathyroid

M 62. Which gland is both an exocrine and
   endocrine gland?
   *   a. pancreas
       b. adrenal
       c. ovary
       d. thyroid
       e. pituitary

E 63. Excess glucose is converted into glycogen
   in the
       a. pancreas.
   *   b. liver.
       c. thymus.
       d. thyroid.
       e. none of these

M 64. Specialized islet cells that secrete hormones
   are found scattered throughout the
       a. adrenal cortex.
       b. liver.
       c. thymus.
       d. adrenal medulla.
   *   e. pancreas.

E 65. Glucagon is produced by the
       a. adrenal cortex.
       b. adrenal medulla.
       c. thyroid.
       d. kidneys.
   *   e. pancreas.

M 66. Insulin directly affects the
       a. secretion of saliva.
       b. storage of proteins.
       c. secretion of pancreatic juices.
   *   d. metabolism of sugar.
       e. utilization of fat reserves.

D 67. If you skip a meal, which of the following
   conditions would prevail?
       a. Insulin levels would rise.
       b. Glucagon levels would rise.
       c. Glycogen would be converted to
          glucose.
       d. Insulin levels would rise; and Glycogen
          would be converted to glucose.
   *   e. Glucagon levels would rise; and
          Glycogen would be converted to
          glucose.

D   68. Which of the following is true of "type 1 diabetes"?
    a. Insulin levels are near normal.
    b. Target cells do not respond to insulin.
    c. It is the more common form of diabetes.
*   d. It is thought to be an autoimmune disease.
    e. It usually occurs in middle-aged people.

M   69. The hormone that is antagonistic in action to glucagon is
    a. norepinephrine.
*   b. insulin.
    c. thyroxine.
    d. epinephrine.
    e. mineralocorticoids.

M   70. The actions of insulin and glucagon could be described as
    a. synergistic.
*   b. antagonistic.
    c. cooperative.
    d. permissive.
    e. mutualistic.

D   71. The pancreatic secretions governing glucose levels are precisely controlled by
    a. neural connections to the pancreas.
    b. the blood-brain barrier.
    c. cooperative interactions.
*   d. homeostatic feedback loops.
    e. releasing factors.

SOME FINAL EXAMPLES OF INTEGRATION AND CONTROL

E   72. Which gland is the remnant of the third eye?
*   a. pineal
    b. pituitary
    c. thyroid
    d. parathyroid
    e. thymus

E   73. The gland that functions in controlling the reproductive cycle is the
    a. thyroid.
*   b. pineal.
    c. thymus.
    d. pancreas.
    e. kidney.

M   74. Which gland is associated with biological clocks or biorhythms?
*   a. pineal
    b. parathyroid
    c. hypothalamus
    d. pituitary
    e. thymus

M   75. Which gland is involved in the maturation of lymphocytes?
    a. thyroid
    b. adrenal
    c. kidney
*   d. thymus
    e. parathyroid

M   76. A group of hormones that are believed to affect the membrane surface receptors of lymphocytes are
*   a. thymosins.
    b. prostaglandins.
    c. erythropoietin.
    d. secretin.
    e. none of these

M   77. Which gland promotes body immune response as its primary function?
    a. pineal
*   b. thymus
    c. thyroid
    d. gonads
    e. adrenal

M   78. A grab-bag of local hormones present in tissues throughout the body (lungs, gut, liver, and the prostate gland in males) that may act as mediators between membrane receptors and the enzymes that activate cAMP are
    a. thymosins.
*   b. prostaglandins.
    c. erythropoietins.
    d. secretins.
    e. all of these

M   79. Prostaglandins
    a. affect blood flow by acting on smooth muscles in the blood cells.
    b. cause menstrual cramps.
    c. may produce strong allergic reactions.
    d. are produced by the corpus luteum if pregnancy does not follow ovulation.
*   e. all of these

D   80. Allergic reactions to airborne dust and pollen may be aggravated by what hormone?
*   a. prostaglandin
     b. somatostatin
     c. insulin
     d. growth hormone
     e. nerve growth factor

M   81. Which of the following diminishes in quantity as a person ages?
     a. epidermal growth factor
     b. prostaglandins
*   c. nerve growth factor
     d. insulin
     e. glucocorticoids

E   82. Which glands secrete pheromones?
*   a. exocrine
     b. ductless
     c. sebaceous
     d. endocrine
     e. digestive

D   83. Pheromones are used primarily for
     a. signaling target cells in the vicinity of the secreting cells.
     b. use as transmitter substances in certain synapses.
*   c. arousing interest in a potential mate.
     d. causing a response in a part of the body some distance from the site of secretion.
     e. all of these, depending on the species

Matching Questions

D   84. Choose the one most appropriate answer for each.

1 _____ adrenal cortex
2 _____ adrenal medulla
3 _____ anterior lobe of pituitary
4 _____ exocrine glands
5 _____ endocrine cells in gastrointestinal tract
6 _____ gonad
7 _____ hypothalamus
8 _____ heart
9 _____ kidneys
10 _____ pancreatic islets
11 _____ parathyroid gland
12 _____ pineal gland
13 _____ posterior lobe of pituitary
14 _____ thymus gland
15 _____ thyroid gland

A. secretes one hormone that increases the metabolic rate and another that inhibits calcium release from bone storage sites

B. secrete insulin and glucagon

C. secretes hormones that prepare accessory reproductive structures for reproduction

D. involved in lymphocyte maturation

E. secretes tropic hormones, growth hormone, and prolactin

F. secretes mineralocorticoids and glucocorticoids

G. secretes a peptide that regulates blood pressure

H. secrete enzymes that help to form angiotensin and erythropoietin

I. participates in reproductive physiology and senses photoperiods

J. secretes (releases into bloodstream) oxytocin and antidiuretic hormone

K. secretes a hormone that promotes calcium release from bone storage sites

L. produces oxytocin and antidiuretic hormone

M. secretes secretin and gastrin

N. secretes epinephrine and norepinephrine

O. secrete pheromones, milk, tears, sweat, and mucus

Chapter 13 The Endocrine System   111

Classification Questions

Answer questions 85–89 in reference to the five endocrine glands listed below:

    a. pituitary
    b. adrenal
    c. pancreas
    d. thyroid
    e. thymus

M   **85.** This gland is the target for corticotropin (ACTH).

E   **86.** Oxytocin is produced in this gland.

M   **87.** Antidiuretic hormone is produced in this gland.

M   **88.** This gland produces a hormone that controls metabolism most directly.

E   **89.** Insulin is produced in this gland.

Answers:
85.	b	86.	a	87.	a
88.	d	89.	c		

Answer questions 90–94 in reference to the five pituitary hormones listed below:

    a. estrogen
    b. luteinizing hormone
    c. somatotropin
    d. oxytocin
    e. antidiuretic hormone

M   **90.** This hormone controls water retention and loss.

M   **91.** The mammary glands are the target for this hormone.

D   **92.** This hormone induces protein synthesis and cell division in young animals.

D   **93.** The kidneys are the target for this hormone.

M   **94.** Uterine contractions are induced by this hormone.

Answers:
90.	e	91.	d	92.	c
93.	e	94.	d		

Answer questions 95–99 in reference to the five endocrine glands listed below:

    a. adrenal cortex
    b. ovary
    c. pineal
    d. thyroid
    e. thymus

D   **95.** This gland controls circadian rhythms.

M   **96.** This gland plays a central role in the immune response.

D   **97.** This gland secretes a hormone that prepares and maintains the uterus for pregnancy.

M   **98.** Progesterone is produced by this organ.

M   **99.** Calcium concentration in the blood is controlled by this gland.

Answers:
95.	c	96.	e	97.	b
98.	b	99.	d		

Selecting the Exception

E   **100.** Four of the five answers listed below are endocrine glands. Select the exception.
    a. thymus gland
*   b. salivary gland
    c. parathyroid gland
    d. thyroid gland
    e. pituitary gland

D   **101.** Four of the five answers listed below are produced by the same lobe of the pituitary. Select the exception.
*   a. antidiuretic hormone
    b. prolactin
    c. corticotropin
    d. somatotropin
    e. luteinizing hormone

M   **102.** Four of the five answers listed below are related by a common source. Select the exception.
    a. glucocorticoids
    b. sex hormones
    c. cortisol
*   d. adrenalin
    e. mineralocorticoid

M   103.   Four of the five answers listed below affect blood glucose level. Select the exception.
    *   a.   calcitonin
        b.   glucagon
        c.   glucocorticoid
        d.   insulin
        e.   epinephrine

D   104.   Four of the five answers listed below directly affect another endocrine gland. Select the exception.
    *   a.   cortisol
        b.   corticotropin
        c.   luteinizing hormone
        d.   follicle-stimulating hormone
        e.   angiotensin

D   105.   Four of the five answers listed below are characteristic of hyperthyroidism. Select the exception.
        a.   excessive weight loss
    *   b.   excessive water loss through urination
        c.   intolerance of heat
        d.   increased heart rate and blood pressure
        e.   excessive sweating

M   106.   Four of the five answers listed below are related by a similar source. Select the exception.
    *   a.   prolactin
        b.   progesterone
        c.   androgen
        d.   estrogen
        e.   testosterone

D   107.   Four of the five answers listed below are related by a similar gland. Select the exception.
        a.   goiter
        b.   deficiency of iodine in the diet
        c.   hypothyroidism
    *   d.   rickets
        e.   excessive stimulation of the thyroid gland

D   108.   Four of the five answers listed below are related by the same action. Select the exception.
        a.   activates vitamin D
        b.   induces resorption of calcium by the kidney
        c.   removes calcium and phosphate from bones
    *   d.   regulates blood volume
        e.   involved in the biofeedback control of extracellular calcium

D   109.   Four of the five answers listed below are characteristic actions of prostaglandins. Select the exception.
        a.   act on smooth muscles in airways
        b.   may produce allergic responses to dust and pollen
        c.   may produce excessive bleeding and cramping during menstrual cycle
    *   d.   generate the fight or flight response
        e.   destroy the corpus luteum if pregnancy does not occur

CHAPTER 14
REPRODUCTIVE SYSTEMS

Multiple-Choice Questions

GIRLS AND BOYS

M   **1.** The major difference between the male and female reproductive systems is the
*   a. provision of a site for fertilization and development in the female.
   b. production of haploid gametes by the male only.
   c. diploid polar bodies of the female.
   d. generation of millions of eggs but only thousands of sperm each month.

M   **2.** The male and female sex organs begin to differentiate during which week of gestation?
   a. fourth week
*   b. seventh week
   c. twelfth week
   d. twentieth week
   e. twenty-eighth week

THE MALE REPRODUCTIVE SYSTEM

E   **3.** In the human male several hundred million sperm are produced by spermatogenesis occurring in
   a. interstitial cells.
   b. the prostate.
*   c. seminiferous tubules.
   d. the ductus deferens.
   e. epididymis.

E   **4.** Sperm are produced in the
*   a. testes.
   b. ductus deferens.
   c. epididymis.
   d. prostate gland.
   e. penis.

M   **5.** Seminal fluid is produced by the
   a. prostate gland.
   b. seminal vesicle.
   c. bulbourethral gland.
   d. urinary bladder.
*   e. all except "urinary bladder."

E   **6.** Which of the following is NOT found in the seminal fluid?
*   a. glucose
   b. buffers
   c. prostaglandins
   d. mucus
   e. fructose

D   **7.** The seminal vesicles ("sperm vessels") are misnamed. Sperm are actually stored in the
   a. ductus deferens.
*   b. epididymis.
   c. prostate.
   d. scrotum.
   e. urethra.

D   **8.** Which of the following is the site where sperm are stored?
   a. ureter
   b. urethra
   c. ductus deferens
   d. vas efferens
*   e. epididymis

D   **9.** If the ductus deferens tubes are cut and tied (vasectomy), the semen will not contain
   a. fructose.
   b. buffers.
   c. mucus.
*   d. sperm.
   e. any of the above

M   **10.** Which of the following is the last structure that a sperm travels through as it leaves the body?
   a. ureter
*   b. urethra
   c. ductus deferens
   d. vas efferens
   e. epididymis

M   **11.** Which of the following is part of the urinary system, not the reproductive system?
*   a. ureter
   b. urethra
   c. ductus deferens
   d. vas efferens
   e. epididymis

MALE REPRODUCTIVE FUNCTION

M   **12.** Which cells are diploid?
- a. spermatids
- b. primary spermatocytes
- c. secondary spermatocytes
- d. spermatogonia
- * e. both primary spermatocytes and spermatogonia

M   **13.** Which cells are produced during meiosis I?
- a. spermatids
- b. primary spermatocytes
- * c. secondary spermatocytes
- d. spermatids
- e. sperm

D   **14.** All but which of the following are the products of meiosis?
- a. male gametes
- b. spermatids
- c. sperm
- d. secondary spermatocytes
- * e. spermatogonia

M   **15.** Sperm become fully motile in the
- a. ductus deferens.
- b. epididymis.
- c. seminiferous tubules.
- * d. vagina.
- e. seminal fluid.

D   **16.** The secretions of the interstitial cells eventually pass into the
- a. semen.
- b. vagina.
- * c. blood.
- d. semen and vagina.
- e. semen, vagina, and blood.

M   **17.** All but which of the following hormones are in some way responsible for the production of sperm?
- a. luteinizing hormone
- b. follicle-stimulating hormone
- c. gonadotropic releasing hormone
- d. testosterone
- * e. human chorionic gonadotropin

M   **18.** Which of the following statements is false?
- * a. Because males do not have a follicle, the follicle-stimulating hormone has no function in males.
- b. The gonads begin differentiation into an ovary or a testis by the seventh week after fertilization.
- c. The testes descend into the scrotal sac before birth.
- d. For sperm to develop properly they must be in an environment that is below body temperature.
- e. The scrotal sac is equipped with muscles that may raise or lower the testes.

M   **19.** Testosterone
- a. stimulates sperm production.
- b. promotes the normal development and maintenance of sexual behavior.
- c. is responsible for secondary sexual characteristics.
- d. is responsible for the development of the male genitalia.
- * e. all of these

D   **20.** Which of the following is NOT involved in a feedback loop in the male reproductive system?
- a. anterior pituitary
- b. hypothalamus
- * c. adrenal gland
- d. Sertoli cells
- e. interstitial cells

D   **21.** The release of testosterone requires
- a. luteinizing hormone.
- b. GnRH.
- c. Sertoli cells.
- * d. luteinizing hormone and GnRH, only.
- e. luteinizing hormone, GnRH, and Sertoli cells.

D   **22.** Based on their response to a particular hormone from the anterior pituitary, Sertoli cells of the male are most like _____ of the female.
- a. uterine cells
- b. oviducts
- * c. ovarian follicles
- d. cervix cells
- e. corpus luteum

D   23. Which of the following structures is NOT
        found in mature sperm?
        a. DNA molecules
        b. acrosome with enzymes
        c. mitochondria
        d. microtubules in the tail
    *   e. ribosomes

THE FEMALE REPRODUCTIVE SYSTEM

E   24. The female reproductive system includes all
        the following EXCEPT
        a. clitoris.
        b. vagina.
        c. oviduct.
        d. ovary.
    *   e. mammary gland.

E   25. The primary reproductive organ in the
        human female is the
        a. uterus.
    *   b. ovary.
        c. vagina.
        d. clitoris.
        e. vulva.

E   26. The passageway that channels ova from the
        ovary into the uterus is known as
        a. a vagina.
        b. a uterus.
    *   c. an oviduct.
        d. an endometrium.
        e. all of these

M   27. The surface of which of the following is
        covered with fingerlike projections that
        produce a sweeping action?
        a. ovary
        b. uterus
        c. vagina
    *   d. oviduct
        e. follicle

D   28. Which of the following is NOT essential to
        the reproductive process?
        a. ovary
        b. oviduct
    *   c. clitoris
        d. vagina
        e. uterus

M   29. Which of the following statements is false?
        a. A female has more oocytes before she
           is born than at any time during her life.
        b. Meiosis II will not occur in an oocyte
           unless it is fertilized.
    *   c. Fertilization occurs in the vagina.
        d. Implantation occurs in the uterus.
        e. The vagina serves as the birth canal.

E   30. The cervix is part of the
        a. vulva.
        b. ovary.
    *   c. uterus.
        d. oviduct.
        e. vagina.

M   31. Which mammal does NOT exhibit seasonal
        sexual activities?
        a. whale
        b. cats
    *   c. primates
        d. horses
        e. dogs

FEMALE REPRODUCTIVE FUNCTION

M   32. Which of the following statements is NOT
        true of the human female?
        a. She produces all the eggs that she ever
           will before she is born.
        b. The process of meiosis may take thirty
           to fifty years to complete.
        c. The primary oocytes lay dormant until
           puberty.
    *   d. She will produce more gametes than
           her male counterpart.
        e. It is possible that more than one egg
           will be released at ovulation.

M   33. Which of the following hormones is
        exclusively female?
        a. follicle-stimulating hormone
        b. luteinizing hormone
    *   c. progesterone
        d. follicle-stimulating hormone and
           progesterone.
        e. follicle-stimulating hormone,
           progesterone, and luteinizing hormone.

M   34. FSH and LH are secreted by the
        a. hypothalamus.
        b. ovaries.
    *   c. anterior pituitary.
        d. testes.
        e. uterus.

M   35. Ovulation is triggered by
     *   a. high levels of LH.
          b. low levels of LH.
          c. high levels of chorionic gonadotropin.
          d. high levels of estrogen.
          e. high levels of progesterone.

M   36. Ovulation involves the
          a. production of the first polar body.
     *   b. release of a secondary oocyte.
          c. beginning of the follicular phase of the menstrual cycle.
          d. suspension of the meiotic process.
          e. deterioration of the corpus luteum.

M   37. Destruction of the corpus luteum, if pregnancy does NOT occur, results from the action of
          a. chorionic gonadotropin.
          b. the luteinizing hormone.
          c. progesterone.
     *   d. prostaglandins.
          e. estrogen.

E   38. Menstrual flow begins in response to
          a. rising levels of FSH and LH.
          b. falling levels of estrogen.
          c. falling levels of progesterone.
     *   d. falling levels of both estrogen and progesterone.

D   39. Which of the following serves to end the menstrual cycle?
          a. a surge in luteinizing hormone
          b. the secretion of human chorionic gonadotropin
     *   c. the secretion of prostaglandins by the corpus luteum that lead to its self-destruction
          d. a rise in the level of progesterone
          e. a drop in the level of gonadotropic hormones in the blood

D   40. Using your knowledge of the feedback loops of human female hormones, which of the following would you predict is the result of high levels of estrogen and progesterone in the blood?
          a. lack of growth of the corpus luteum
     *   b. absence of monthly ovulation
          c. increased secretion of FSH
          d. increased levels of LH
          e. all of these

VISUAL SUMMARY OF THE MENSTRUAL CYCLE

D   41. The menstrual flow is the result of
          a. no implantation of a zygote. blastocyst
          b. decreased levels of progesterone.
          c. discarded uterine linings. — blastocyst
          d. no implantation of a zygote and discarded uterine linings, only.
     *   e. no implantation of a zygote, discarded uterine linings, and decreased levels of progesterone. blastocyst

E   42. Menstrual flow results in the discharge of
          a. the follicle.
          b. the corpus luteum.
     *   c. the endometrial lining.
          d. surface cells from the vagina.
          e. blood from the blood vessels on the outer surface of the uterus.

E   43. Ovulation is triggered primarily by
     *   a. a surge of LH that occurs halfway through the menstrual cycle.
          b. the falling levels of estrogen and progesterone.
          c. the rising levels of progesterone.
          d. both a surge of LH that occurs halfway through the menstrual cycle and the rising levels of progesterone.

M   44. Ovulation is triggered by a surge in the level of _____ in the circulatory system.
          a. estrogen
          b. follicle-stimulating hormone
     *   c. luteinizing hormone
          d. progesterone
          e. human chorionic gonadotropin

SEXUAL INTERCOURSE AND FERTILIZATION

M   45. Orgasm is necessary for
     *   a. ejaculation of semen.
          b. pregnancy.
          c. erection.
          d. sexual arousal.
          e. all of these

E   **46.** The union of egg and sperm is called
   a. erection.
   b. coitus.
   c. orgasm.
 * d. fertilization.
   e. menarche.

E   **47.** Sexual intercourse is technically referred to as
   a. erection.
 * b. coitus.
   c. orgasm.
   d. fertilization.
   e. menarche.

CONTROL OF FERTILITY

E   **48.** Which of the following is the most effective in helping prevent venereal disease?
 * a. condoms
   b. the Pill
   c. douching
   d. the IUD
   e. rhythm

D   **49.** Which of the following is the most effective contraceptive approach of those listed?
   a. withdrawal
 * b. diaphragm
   c. cheap condoms
   d. douche
   e. rhythm method

E   **50.** The type of contraception that functions by preventing implantation (not fertilization) is
   a. tubal ligation.
   b. spermicidal jelly.
   c. birth control pills.
 * d. intrauterine device (IUD).
   e. diaphragm.

E   **51.** The type of contraception that works because ovulation is prevented is
   a. rhythm.
   b. tubal ligation.
 * c. birth control pills.
   d. diaphragm.
   e. condom.

M   **52.** Which of the following is NOT an effective temporary method of contraception?
   a. birth-control pills
 * b. vasectomy
   c. condoms
   d. IUD plus spermicide
   e. spermicidal jelly or foam

E   **53.** Of the following, which is the least successful method of birth control?
   a. early withdrawal
   b. a condom alone
   c. a spermicidal jelly or foam alone
 * d. douching
   e. the Pill

M   **54.** A contraceptive pill contains
   a. estrogen.
   b. progesterone.
   c. the follicle-stimulating hormone.
   d. the luteinizing hormone.
 * e. both estrogen and progesterone.

COPING WITH INFERTILITY

D   **55.** In *in vitro* fertilization, it is possible to tell that an ovum has been fertilized by the presence of
   a. human chorionic gonadotropin.
   b. progesterone.
   c. estrogen.
   d. testosterone.
 * e. the second polar body.

M   **56.** The name of the procedure in which fertilization is accomplished in a laboratory dish and the resulting zygote is placed in the women's oviducts is called
   a. embryo transfer.
 * b. ZIFT.
   c. AID.
   d. GIFT.
   e. in vivo fertilization.

Matching Questions

D   57. Matching I. Choose the one most appropriate answer for each.

1 _____ abortion
2 _____ abstention
3 _____ coitus
4 _____ douching
5 _____ ejaculation
6 _____ implantation
7 _____ lactation
8 _____ menopause
9 _____ menstruation
10 _____ miscarriage
11 _____ orgasm
12 _____ ovulation
13 _____ tubal ligation
14 _____ vasectomy

A. the production and secretion of milk
B. sloughing off of endometrium stops permanently
C. birth control exercised after development begins
D. an abortion that occurs spontaneously
E. the burrowing of the blastocyst into the uterus
F. the release of an egg from the ovary
G. the release of seminal fluid from the male reproductive tract
H. a 100 percent effective method of preventing conception
I. a highly unreliable form of birth control
J. the periodic elimination of the uterine lining
K. characterized by involuntary muscle contractions, release of tension, and warmth
L. sexual intercourse
M. sperm and egg cannot meet because section of oviduct is missing
N. cutting and tying of ductus deferens

Answers:

1. C	2. H	3. L
4. I	5. G	6. E
7. A	8. B	9. J
10. D	11. K	12. F
13. M	14. N	

D   58. Matching II. Choose the one most appropriate answer for each.

1 _____ acrosome
2 _____ GIFT
3 _____ menarche
4 _____ cervix
5 _____ clitoris
6 _____ endometrium
7 _____ epididymis
8 _____ FSH
9 _____ Leydig cells of testis
10 _____ LH
11 _____ placenta
12 _____ seminal vesicles
13 _____ ductus deferens

A. in females, acts on ruptured follicle to produce corpus luteum
B. structures that secrete mucus and nutrients absorbable by sperm; open into the ejaculatory duct
C. transfer of gametes to oviduct by lab technician
D. two of these connect seminiferous tubules with ductus deferens
E. opening between uterus and vagina
F. organ that supplies the embryo/fetus with nutrients and removes waste products
G. connect epididymides with ejaculatory duct
H. cap over the head of a sperm; contains lytic enzymes that help penetrate egg membrane
I. first menstruation
J. testosterone produced here
K. part of vulva; develops from same embryonic tissues as does the penis in males
L. the uterine lining
M. acts on gonad to help mature gametes; released from anterior lobe of pituitary

Answers:

1. H	2. C	3. I
4. E	5. K	6. L
7. D	8. M	9. J
10. A	11. F	12. B
13. G		

Classification Questions

Answer questions 59–63 in reference to the five stages of sperm development listed below:

    a. spermatogonia
    b. secondary spermatocyte
    c. primary spermatocyte
    d. spermatid
    e. sperm

D   **59.** Stage of development in which the male sex cell is first in the haploid condition.

M   **60.** These continue to undergo mitosis throughout the reproductive life of the male.

D   **61.** This represents the product of the first meiotic division.

D   **62.** This is a mitotic product, but then undergoes meiosis.

M   **63.** At this stage of development the male sex cell is fully motile.

Answers: 59. b   60. a   61. b
             62. c   63. e

Answer questions 64–68 in reference to the five stages and structures involved in the development of the human egg:

    a. oogonium
    b. primary oocyte
    c. secondary oocyte
    d. polar body
    e. secondary follicle

M   **64.** This structure becomes the mature egg only after fertilization is begun.

D   **65.** This structure contains the first sex cell stage to be in the haploid state.

M   **66.** This structure contains a full haploid set of chromosomes, but will never be fertilized.

M   **67.** This structure divides by mitosis during the fetal stages of development until all of the approximately 2 million potential eggs have been formed.

D   **68.** This stage lies dormant between birth and puberty.

Answers: 64. c   65. c   66. d
             67. a   68. b

Selecting the Exception

M   **69.** Three of the four answers listed below produce portions of the seminal fluid. Select the exception.
   *   a. epididymis
         b. prostate
         c. seminal vesicle
         d. bulbourethral gland

M   **70.** Three of the four answers listed below are related by the number of chromosomes present. Select the exception.
         a. sperm
   *   b. spermatogonia
         c. secondary spermatocyte
         d. spermatids

E   **71.** Three of the four answers listed below are all parts of a sperm. Select the exception.
         a. flagella
         b. midpiece
         c. acrosome
   *   d. polar body

D   **72.** Four of the five answers listed below are related by a common quantity. Select the exception.
   *   a. urethra
         b. testis
         c. ejaculatory duct
         d. ductus deferens
         e. epididymis

D   **73.** Four of the five answers listed below are true of testosterone. Select the exception.
         a. promotes secondary sex characteristics
         b. controls sexual behavior
         c. necessary for growth and function of male reproductive tract
         d. stimulates spermatogenesis
   *   e. produced by spermatogonia cells

E   **74.** Four of the five answers listed below are related by gender. Select the exception.
   *   a. Sertoli cells
         b. cervix
         c. clitoris
         d. myometrium
         e. vulva

M   **75.** Four of the five answers listed below are
        related by a common location. Select the
        exception.
          a. follicle
          b. corpus luteum
    *   c. cervix
          d. oogonium
          e. primary oocyte

M   **76.** Four of the five answers listed below are
        related by a similar effectiveness. Select the
        exception.
          a. pill
          b. tubal ligation
          c. vasectomy
    *   d. douching
          e. IUD

CHAPTER 15
DEVELOPMENT AND AGING

Multiple-Choice Questions

STAGES OF DEVELOPMENT

M   **1.** At the end of gastrulation, which of the following are produced?
    a. hollow balls of cells
  * b. embryos with germ layers
    c. solid balls of cells
    d. maternal messages
    e. all of these

E   **2.** Which stage in development occurs first?
    a. cleavage
    b. morula
    c. gastrula
  * d. zygote
    e. blastula

E   **3.** Which of the following is a single-layered, hollow ball of cells?
    a. cleavage
    b. morula
    c. gastrula
    d. zygote
  * e. blastula

E   **4.** The germ layers are formed in which of the following stages?
    a. cleavage
    b. morula
  * c. gastrula
    d. zygote
    e. blastula

D   **5.** Select the correct sequence of animal developmental events.
    a. fertilization >>> cleavage >>> gastrula >>> blastula
  * b. fertilization >>> cleavage >>> blastula >>> gastrula
    c. fertilization >>> blastula >>> cleavage >>> gastrula
    d. fertilization >>> gastrula >>> blastula >>> cleavage

E   **6.** The heart, muscles, bones, and blood develop primarily from
    a. ectoderm.
  * b. mesoderm.
    c. endoderm.
    d. the placenta.
    e. the gray crescent.

M   **7.** If an experimenter interferes with the mesoderm in an egg, which of the following systems would NOT be affected by the experimenter?
    a. circulatory system
    b. muscular system
    c. reproductive system
  * d. integumentary system
    e. excretory system

E   **8.** The process of cleavage most commonly produces a
    a. zygote.
  * b. morula.
    c. gastrula.
    d. puff.
    e. third germ layer.

M   **9.** Which embryonic tissue is incorrectly associated with its derivative?
  * a. skin from mesoderm
    b. nervous system from ectoderm
    c. liver from endoderm
    d. circulatory system from mesoderm

E   **10.** Muscles differentiate from which tissue?
    a. ectoderm
    b. endoderm
  * c. mesoderm
    d. all of these

E   **11.** Shortly after fertilization, successive cell divisions convert the zygote into a multicellular embryo during a process known as
    a. meiosis.
    b. parthenogenesis.
    c. embryonic induction.
  * d. cleavage.
    e. invagination.

M **12.** The mesoderm is responsible for the formation of all of the following adult tissues EXCEPT
    a. reproductive system.
    b. circulatory system.
*   c. nervous system.
    d. muscle system.
    e. excretory system.

E **13.** In the following list of developmental events, which occurs last?
*   a. tissue specialization
    b. gamete formation
    c. gastrulation
    d. cleavage
    e. organ formation

D **14.** In the human, which of the following events would occur over the longest period of time?
    a. sperm production
    b. cleavage
    c. fertilization
    d. gastrulation
*   e. growth and tissue specialization

M **15.** During which of the following stages do cells of identical genetic makeup become structurally and functionally different from one another according to the genetically controlled developmental program of the species?
    a. cleavage
*   b. differentiation
    c. morphogenesis
    d. metamorphosis
    e. ovulation

M **16.** In the process of differentiation,
    a. some daughter cells usually receive varying assortments of genes.
*   b. cells with identical assortments of genes come to have different individual genes expressed.
    c. cells become specialized as a result of meiosis.
    d. daughter cells acquire different characteristics as a result of mutations that have occurred.
    e. all of these

M **17.** During which of the following do sets of cells become structurally and functionally unique and arranged into tissues and organs?
    a. cleavage
*   b. organogenesis
    c. morphogenesis
    d. metamorphosis
    e. ovulation

D **18.** The explanation for the differences between a cell from the human liver and a cell from the skin is the
    a. maternal and paternal origins of the cell types.
    b. gene content of the two cells.
*   c. expression of the genes in the two cells.
    d. fact that they are in different parts of the body.

THE BEGINNINGS OF YOU—EARLY EVENTS IN DEVELOPMENT

E **19.** Fertilization in mammals occurs in the
    a. ovary.
    b. uterus.
    c. vagina.
*   d. oviduct.
    e. follicle.

E **20.** The average number of sperm that are deposited in the vagina during an ejaculation is between
    a. 150,000 and 350,000.
    b. 1.5 and 3.5 million.
    c. 15 and 35 million.
*   d. 150 and 350 million.
    e. 1.5 and 3.5 billion.

E **21.** The process wherein the sperm become ready to fertilize the egg is known as
    a. differentiation.
    b. cleavage.
*   c. capacitation.
    d. morphogenesis.
    e. implantation.

M **22.** The signal for the second meiotic division in the egg is
    a. implantation in the uterus.
*   b. entry of the sperm into the egg.
    c. capacitation of the sperm.
    d. fusion of the sperm nucleus with the egg nucleus.
    e. cleavage of the zygote.

M 23. The first several cleavages after fertilization occur in the
   a. uterus.
   b. ovary.
   c. vagina.
 * d. oviduct.
   e. any of the above except vagina

D 24. In the process of blastocyst formation,
   a. the size of individual cells decreases.
   b. the number of cells increases.
   c. the total amount of cytoplasm remains about the same.
   d. only the size of individual cells decreases and the number of cells increases.
 * e. the size of individual cells decreases, the number of cells increases, and the total amount of cytoplasm remains about the same.

M 25. Which of the following is NOT true of identical twins?
   a. They are appropriately called monozygotic twins.
   b. They always are the same sex.
 * c. They occur more frequently than fraternal twins.
   d. They have exactly the same genotypes and could appropriately be called clones.

M 26. During a human pregnancy, implantation occurs at which stage?
   a. zygote
   b. early cleavage
 * c. blastocyst
   d. gastrula
   e. morula

E 27. Implantation occurs in the
   a. ovary.
 * b. uterus.
   c. vagina.
   d. oviduct.
   e. follicle.

M 28. Which of the following statements is false?
   a. Ovulation occurs when the follicle ruptures and releases an egg.
   b. Cleavage is the divison of the zygote.
   c. Fertilization occurs in the upper regions of the oviduct.
   d. The blastocyst implants in the endometrial lining of the uterus.
 * e. Implantation occurs about 36 hours after fertilization.

E 29. During human development, which of the following gives rise to the embryo?
   a. trophoblast
   b. amnion
 * c. inner cell mass
   d. chorion
   e. placenta

E 30. Which of the following is responsible for the actual invasion of the lining of the uterus?
   a. trophoblast
   b. amnion
   c. inner cell mass
 * d. blastocyst
   e. placenta

E 31. In an ectopic pregnancy, implantation occurs
   a. in the oviduct.
   b. on the outside of the uterus.
   c. on the abdominal wall.
   d. on the surface of the ovary.
 * e. any of the these

M 32. The presence of which hormone in a mother's urine indicates that she is pregnant?
   a. luteinizing hormone
   b. follicle-stimulating hormone
 * c. chorionic gonadotropin
   d. progesterone
   e. estrogen

M 33. Which of the following hormones is produced only when a woman is pregnant?
   a. testosterone
   b. gonadotropic-releasing hormone
 * c. human chorionic gonadotropin
   d. estrogen
   e. progesterone

M 34. The fact that a blastocyst has been implanted on the uterine wall can be demonstrated by the presence of
 * a. human chorionic gonadotropin.
   b. progesterone.
   c. estrogen.
   d. testosterone.
   e. second polar body.

EXTRAEMBRYONIC MEMBRANES

M 35. In humans, the fluid immediately
        surrounding the embryo is contained in the
        a. allantois.
        b. placenta.
        c. chorion.
     * d. amnion.
        e. yolk sac.

M 36. The outermost membrane that forms the
        majority of the placenta is the
        a. amnion.
        b. allantois.
     * c. chorion.
        d. yolk sac.
        e. umbilical cord.

M 37. Which of the following membranes is
        associated with the formation of the
        placenta?
        a. amnion
        b. yolk sac
     * c. chorion
        d. allantois
        e. none of these

M 38. Which of the following would be expected
        to diffuse in greater amounts from the fetal
        blood to the mother's blood?
        a. oxygen
     * b. urea
        c. hormones
        d. antibodies
        e. nutrients

THE MAKING OF AN EARLY EMBRYO—A CLOSER LOOK

M 39. The first evidence of where the nerve cord
        will be in embryo is the
        a. neural tube.
        b. chorionic villus.
        c. embryonic somite.
     * d. primitive streak.
        e. neural plate.

M 40. Which of the following systems is the first
        of those listed to begin development in the
        human embryo?
     * a. nervous system
        b. excretory system
        c. reproductive system
        d. skeletal system
        e. endocrine system

M 41. What is the name of the birth defect in
        which the neural tube fails to close
        completely?
        a. neurulation
        b. fossa ovalis
        c. organogenesis
     * d. spina bifida
        e. capacitation

EMERGENCE OF DISTINCTLY HUMAN FEATURES

M 42. A embryo with two X chromosomes will
        develop female characters because
        a. the presence of female hormones in
            development.
     * b. the absence of male hormones during
            early embryology.
        c. of the SRY regions of the X
            chromosomes.
        d. of maternal substances passed to the
            zygote during fertilization.
        e. of human chorionic gonadotropin.

E 43. The embryo is recognizable as human and is
        called a fetus by which week of pregnancy?
     * a. eighth
        b. twelfth
        c. sixteenth
        d. twentieth
        e. twenty-four

FETAL DEVELOPMENT

M 44. All of the following are true about the
        umbilical arteries EXCEPT which one?
        a. There are two of them.
        b. They carry oxygen-poor blood.
        c. They take blood from the fetus to the
            placenta.
        d. They are housed in the umbilical cord.
     * e. They carry blood that is rich in
            nutrients.

E 45. What is the name given to the opening that
        allows blood to flow from the right atrium
        directly into the left atrium of the fetal
        heart?
        a. ductus arteriosus
     * b. foramen ovale
        c. vernix caseosa
        d. ductus venosus
        e. lanugo

E   46. Which of the following is NOT a structure associated in some way with the bypassing of blood around the fetal lungs?
   a. ductus arteriosus
   b. fossa ovalis
 * c. ductus venosus
   d. foramen ovale

D   47. Which of the following does NOT occur in the first trimester?
   a. formation of a heart
   b. disappearance of the tail
   c. formation of internal organs
 * d. detection of movement of the fetus
   e. segmentation and development of somites

E   48. The soft, fuzzy hair covering the body of a fetus in the second trimester is called
   a. vernix caseosa.
   b. fossa ovalis.
   c. chorionic villus.
 * d. lanugo.
   e. colostrum.

M   49. Respiratory distress syndrome is due to
   a. a lack of surfactant in the lungs.
   b. premature birth.
   c. lack of prenatal care.
 * d. a lack of surfactant in the lungs and premature birth.
   e. premature birth and lack of prenatal care.

BIRTH AND BEYOND

D   50. The hormones that are most directly involved with the process of birth are
   a. estrogen and oxytocin.
 * b. oxytocin and prostaglandins.
   c. oxytocin and progesterone.
   d. prolactin and estrogen.
   e. prostaglandin and progesterone.

M   51. Which of the following occurs in all three stages of the birthing process—"labor?"
   a. dilation
   b. delivery
 * c. contractions
   d. afterbirth
   e. rupture of amnion

M   52. The female hormones that participate in milk production for the newborn are
 * a. prolactin and oxytocin.
   b. prolactin and estrogen.
   c. prolactin and progesterone.
   d. oxytocin and estrogen.
   e. oxytocin and progesterone.

M   53. Milk production in women is stimulated by
   a. estrogen.
 * b. prolactin.
   c. oxytocin.
   d. prostaglandin.
   e. progesterone.

HOW THE MOTHER'S LIFESTYLE AFFECTS EARLY DEVELOPMENT

M   54. Which disease may produce a malformed embryo if the mother develops the disease early in pregnancy?
 * a. German measles
   b. chicken pox
   c. red measles
   d. hepatitis
   e. mumps

E   55. Which drug, if taken during pregnancy, results in infants without arms or legs?
   a. tetracycline
 * b. thalidomide
   c. streptomycin
   d. salicylic acid
   e. codeine

M   56. Exposure to which of the following would cause a newborn to experience improper nervous system development, chronic irritability, oxygen deprivation and long-term emotional problems?
 * a. cocaine
   b. Retin-A
   c. thalidomide
   d. streptomycin
   e. alcohol

SUMMARY OF DEVELOPMENTAL STAGES

E   57. Which of the following is true concerning postnatal development?
     a. Infancy is defined as the first two years of life after birth.
     b. Senescence results in increased hormone secretion.
     c. Postnatal development occurs before birth.
  * d. In general boys experience their pre-adolescent growth spurt later than girls.
     e. Early childhood is a time of rapid cell division in nerve cells.

E   58. The process of aging is described by the term
  * a. senescence.
     b. puberty.
     c. neonate.
     d. morulation.
     e. fetalism.

WHY DO WE AGE?

E   59. Limited division potential is one of the explanations for
     a. puberty.
     b. chronic irritability in some babies exposed to drugs.
  * c. the aging process.
     d. the growth spurts that are typical of pre-adolescents.
     e. mental retardation in poorly nourished fetuses.

E   60. Some researchers believe the many of the changes seen in aging tissues could result from damage by _____ oxygen.
     a. isotopes of
     b. too much
  * c. free radicals of
     d. the lack of
     e. reduced

AGING OF SKIN, MUSCLE, THE SKELETON, AND INTERNAL TRANSPORT SYSTEMS

E   61. Which of the following is NOT one of the signs of aging in the skin?
     a. increased collagen
     b. reduced fibroblasts
  * c. increase in sweat glands
     d. less pigmentation in hair
     e. increased fat deposits

E   62. As bones age in older persons, they
     a. have increased calcium deposits.
     b. become stronger.
     c. are less porous.
  * d. more brittle.
     e. increase in length.

E   63. Which of the following is NOT true of aging cardiovascular systems?
     a. Pumping ability diminishes.
     b. Less blood and oxygen are delivered to the tissues.
  * c. Blood vessel walls become more elastic.
     d. Plaque deposits increase on vessel walls.
     e. Heart size usually becomes smaller.

AGE-RELATED CHANGES IN SOME OTHER BODY SYSTEMS

M   64. Clumps of microtubules in aging neurons are called
     a. beta amyloid.
  * b. neurofibrillary tangles.
     c. neural atrophy.
     d. plaques.
     e. ganglionic fossae.

E   65. All of the following are usually a part of the aging reproductive systems EXCEPT
     a. declining levels of estrogen.
     b. menopause.
  * c. sexual response.
     d. decreased fertility.
     e. declining testosterone levels.

Classification Questions

Answer questions 66–70 in reference to the five stages of development listed below:

    a. zygote
    b. blastocyst
    c. morula
    d. gastrula
    e. embryo

M   **66.** This stage appears as a multicellular, hollow sphere.

E   **67.** This is the fertilized egg.

M   **68.** This stage might be described as a "solid ball."

D   **69.** The gut cavity of an animal forms during this stage.

D   **70.** The major germ layers are formed during this stage.

Answers: 66. b   67. a   68. c
                69. d   70. d

Selecting the Exception

M   **71.** Four of the five answers listed below are events occurring after fertilization. Select the exception.
        a. cleavage
*   b. gametogenesis
        c. blastula
        d. gastrulation
        e. organogenesis

M   **72.** Four of the five answers listed below are produced by the same germ layer. Select the exception.
*   a. nervous system
        b. muscle system
        c. circulatory system
        d. reproductive system
        e. excretory system

D   **73.** Four of the five answers listed below are related by a matching feature. Select the exception.
*   a. blastocyst
        b. amnion
        c. allantois
        d. yolk sac
        e. chorion

M   **74.** Four of the five answers listed below are related by a common theme. Select the exception.
        a. alcohol
        b. thalidomide
        c. German measles
*   d. vitamins
        e. antibiotics

M   **75.** Four of the five answers listed below are events associated with the aging process. Select the exception.
        a. skin wrinkling
*   b. increased metabolism
        c. decreased muscle mass
        d. altered collagen
        e. increased fat deposition

CHAPTER 16

LIFE AT RISK: INFECTIOUS DISEASE

Multiple-Choice Questions

VIRUSES AND INFECTIOUS PROTEINS

E **1.** Which of the following could be called "pathogens"?
- a. viruses
- b. bacteria
- c. protozoans
- d. bacteria and protozoans only, because they are alive
- * e. viruses, bacteria, and protozoans.

E **2.** Which statement is NOT true?
- a. Viruses are not able to move by themselves.
- b. Viruses are not able to reproduce by themselves.
- * c. Viruses are not structurally organized.
- d. Some biologists consider that viruses are forms of life and other biologists consider them to be nonlife.
- e. Viruses contain instructions to manufacture themselves.

E **3.** When a virus takes over the machinery of a cell, it forces the cell to manufacture
- a. more mitochondria for energy for the virus.
- b. more liposomes to isolate themselves from water.
- c. more food particles.
- * d. more viral particles.
- e. more Golgi bodies so that the cell will secrete the excess viruses.

M **4.** Most scientists do not consider viruses to be "alive" because
- a. they have no genes.
- * b. their metabolic machinery is borrowed from the host cell.
- c. they are unable to reproduce.
- d. no definite structural features are seen under the microscope.
- e. all of these

M **5.** Which of the following is false?
- * a. The capsids of all viruses are alike.
- b. The virus uses either DNA or RNA as its core, but not both.
- c. Viruses can be replicated only after they enter a living cell.
- d. Most viruses have a protein coat or covering.
- e. A virus may not kill a host cell but may become inactive for a period of latency.

D **6.** In viral replication, all but which one of the following occur before capsid formation?
- a. attachment to host cell
- b. nucleic acid replication
- * c. release of new viral particles
- d. injection of viral nucleic acid into cell

D **7.** Latency in viruses can be associated with all but which one of the following?
- a. replication
- b. reverse transcriptase
- c. gene integration
- * d. rapid production of viral particles
- e. retroviruses

D **8.** All but which one of the following are true of retroviruses?
- * a. Viral RNA becomes integrated into host genome.
- b. The life cycle can involve latency.
- c. They infect animal cells.
- d. They are responsible for AIDS infection.
- e. Transcriptase enzymes are coded for by viral genes.

M **9.** Retroviruses
- a. can infect animal cells.
- b. can have periods of latency.
- c. use reverse transcriptase.
- d. cause RNA to be transcribed to DNA.
- * e. all of these

D **10.** Which virus is an RNA virus?
- a. adenovirus
- * b. retrovirus
- c. parvovirus
- d. Herpes virus
- e. papovavirus

M 11. A virus is characterized by all of the following EXCEPT
   *   a. enzymes of respiration.
        b. nucleic acid core.
        c. noncellular organization.
        d. protein coat.
        e. capsid.

M 12. Retroviruses are characterized by
        a. an RNA core.
        b. latency in pathways of replication.
        c. the enzyme reverse transcriptase.
        d. being the causative agent for AIDS.
   *   e. all of these

E 13. Infective proteins are known as
        a. retroviruses
        b. vivoids.
        c. viruses.
   *   d. prions.
        e. none of these because nucleic acids are needed for infections.

BACTERIA—THE UNSEEN MULTITUDES

E 14. Spherical bacteria are called
        a. bacilli.
        b. spirilla.
   *   c. cocci.
        d. bacteriophages.
        e. all of these

M 15. A helical or spiral bacterium is called a
   *   a. spirillum.
        b. bacillus.
        c. coccus.
        d. bacillus or coccus.

M 16. Peptidoglycan is
        a. found in the chromosomes of most bacteria.
   *   b. composed of polysaccharides crosslinked with proteins.
        c. composed of long polypeptides held together by disulfide bridges.
        d. a unique combination of protein lipid and fat.

M 17. Which statement about bacteria is true?
        a. They are diploid organisms.
        b. They produce gametes.
   *   c. They possess circular DNA molecules.
        d. They are eukaryotic.

E 18. In bacteria, DNA is found
        a. in the nucleus alone.
        b. in organelles alone.
        c. in both the nucleus and organelles.
   *   d. as a single circular thread, and possibly as additional pieces.
        e. as particles scattered throughout the entire bacterial cell.

M 19. Which of the following statements is false? Antibiotics
        a. serve as an agent of natural selection in pathogenic bacteria.
   *   b. are effective against viruses.
        c. may produce potent side effects.
        d. are normal metabolic by-products of certain microorganisms.
        e. when used by women often have to be accompanied by antifungal drugs to control yeast infections.

M 20. Antibiotic resistance is the result of
        a. mutation in the pathogenic organism.
   *   b. natural selection of the best adapted organisms.
        c. using too low a dose of antibiotic.
        d. genetic engineering.
        e. increased pressure from viruses.

E 21. Antibiotic drugs are most effective against
   *   a. bacteria primarily.
        b. viruses primarily.
        c. mostly bacteria and some viruses.
        d. bacteria and viruses, equally.
        e. all pathogens.

INFECTIOUS PROTOZOA AND WORMS

E 22. Unlike bacteria, protozoa are
        a. single-celled.
   *   b. eukaryotes.
        c. multicellular.
        d. capable of causing disease.
        e. photosynthetic.

M 23. All of the following cause severe diarrhea as a symptom EXCEPT:
        a. *Entamoeba*
        b. giardiasis
        c. *Cryptosporidium*
   *   d. African trypanosomiasis

M 24. Which of the following is not a true worm?
   a. pinworm
   b. hookworm
 * c. ringworm
   d. roundworm
   e. tapeworm

TRANSMISSION AND PATTERNS OF INFECTIOUS DISEASE

M 25. What would be the most common method of transmission for malaria?
   a. direct contact with the pathogen
   b. inhaling the pathogen
 * c. encounter with a mosquito vector
   d. indirect with contaminated food

E 26. Diseases occurring more or less continuously but in rather localized populations describes
   a. sporadic.
 * b. endemic.
   c. epidemic.
   d. pandemic.
   e. virulent.

M 27. All of the following are viral diseases EXCEPT
   a. warts.
   b. herpes.
 * c. gonorrhea.
   d. common cold.
   e. AIDS.

E 28. Bacteria cause disease by
   a. taking over a cell's machinery for making new DNA.
   b. punching holes in the host cell's membrane.
 * c. releasing toxins into the bloodstream.
   d. inhibiting DNA replication.
   e. all of these.

THE HUMAN IMMUNODEFICIENCY VIRUS AND AIDS

E 29. The reason AIDS is so serious is that
   a. the excessive immune reaction leads to death.
   b. it is so highly contagious.
 * c. it is fatal.
   d. it is caused by a retrovirus.
   e. many natural reservoirs may spread the disease at any time.

E 30. The greatest percent of HIV victims live in
   a. the United States.
   b. Europe.
   c. South American countries.
 * d. Africa.
   e. Asia.

M 31. One of the possible origins of the HIV virus is a mutation from a similar virus in
   a. contaminated fish.
 * b. chimpanzees.
   c. bacteria.
   d. hemophiliacs.
   e. Kaposi's sarcoma.

E 32. The HIV virus was identified in
 * a. early 1980s..
   b. 1959.
   c. the early 1930's.
   d. 1945.

M 33. Which of the following statements is false?
   a. The virus that causes AIDS is the human immunodeficiency virus.
   b. The AIDS virus is a retrovirus.
   c. The AIDS virus attacks macrophages and helper T cells.
   d. Even though the AIDS virus does not have its own DNA, it causes the host to produce DNA that then becomes incorporated into host chromosomes.
 * e. Antibodies to the AIDS virus can be detected immediately after infection occurs.

M 34. The term 'provirus' describes
   a. RNA that has become integrated into host cell DNA.
   b. a virus before it is capable of causing an infection.
   c. a virus that has just entered a host cell.
 * d. viral DNA that has become integrated into the host's genetic material.
   e. early evolutionary versions of present-day intracellular parasites.

E 35. The human immunodeficiency virus (HIV) primarily destroys which cells?
   a. B lymphocytes
   b. brain cells
   c. erythrocytes
 * d. CD4 lymphocytes
   e. killer T

E   36. Kaposi's sarcoma is characteristic of people who have
  * a. AIDS.
    b. allergic reactions.
    c. a hypersensitive immune system.
    d. ancestors who come from Cyprus.
    e. herpes.

M   37. Which of the following statements is true concerning HIV infection?
    a. There are no clinical symptoms in the initial infection.
  * b. It may take up to three years before antibodies to HIV appear.
    c. During the "latent" period, the immune system is destroying HIV as quickly as it is being produced.
    d. HIV kills its victims by destroying the liver cells.
    e. Full-blown AIDS is characterized by a greater than normal supply of T cells.

TREATING AND PREVENTING HIV INFECTION AND AIDS

E   38. Of the following, AIDS is most likely to be transferred by
    a. casual contact.
    b. food.
    c. water.
  * d. sexual intercourse.
    e. insect bites.

E   39. Which one of the following does NOT routinely contain enough HIV for successful transmission?
    a. blood
  * b. saliva
    c. semen
    d. vaginal secretions
    e. breast milk

D   40. Which of the following is NOT under investigation as a method to prevent HIV?
    a. blocking HIV protease
    b. vaccines
  * c. increasing the production of macrophages
    d. disable proteins needed for replication
    e. compounds that disrupt the binding of HIV to host cells

D   41. Only one of the following is a major hindrance to the development of a vaccine against HIV. Which one is it?
    a. The HIV has not been identified and characterized.
    b. The mechanism of HIV replication is unknown.
    c. HIV does not mutate.
  * d. HIV integrates into host cell DNA away from antibodies.
    e. The techniques of genetic engineering necessary to produce vaccines have not been perfected.

E   42. Practical prevention of the spread of AIDS depends on
    a. covering the respiratory passages to block inhalation of virus.
  * b. behavior modification to minimize risk.
    c. faithfully taking anti-viral drugs.
    d. quarantining persons who are infected.
    e. immediate vaccination of all persons who are in the company of infected persons.

A TRIO OF COMMON STDs

M   43. Which of the following is largely asymptomatic in females?
    a. pubic lice
    b. hepatitis B
  * c. gonorrhea
    d. syphilis
    e. HIV

M   44. Which of these is characterized by various stages in the progression of the disease including lengthy latent periods?
    a. pelvic inflammatory disease
    b. scabies
    c. gonorrhea
  * d. syphilis
    e. genital warts

E   45. Which of the following is technically NOT a sexually transmitted disease but rather a condition that can result from infection by STD organisms?
    a. chlamydia
  * b. pelvic inflammatory disease
    c. genital herpes
    d. chancroid
    e. genital warts

A ROGUE'S GALLERY OF VIRAL STDs AND OTHERS

E   46. Which of the following can NOT be
          successfully treated with antibiotics?
  *   a. genital herpes
          b. gonorrhea
          c. syphilis
          d. chancroid
          e. chlamydia

E   47. Which of the following CANNOT be cured
          by drug treatment?
  *   a. scabies
          b. candidiasis
          c. *Trichomonas*
          d. syphilis
          e. hepatitis B

E   48. Vaccination is available against
          a. Herpes simplex viruses
  *   b. hepatitis B
          c. gonorrhea
          d. a and b
          e. b and c

M   49. Which of the following would NOT display
          visible symptoms on the external genitals?
          a. pelvic inflammatory disease
          b. vaginitis
          c. hepatitis B
  *   d. a and c
          e. a, b , and c

E   50. All of the following are caused by a
          bacterium EXCEPT
          a. gonorrhea
          b. syphilis
          c. nongonococcal urethritis
  *   d. scabies
          e. chlamydia

M   51. Which of the following is caused by a
          yeast?
          a. scabies.
  *   b. candidiasis.
          c. "crabs".
          d. nongonococcal urethritis.
          e. genital herpes.

Matching Questions

D   52. Match the letters of the definitions on the
          right with the appropriate numbered terms
          on the left.

          1 ____ HIV
          2 ____ AZT
          3 ____ *Treponema*
          4 ____ genital warts
          5 ____ *Trichomonas*
          6 ____ *Candida*
          7 ____ chancre
          8 ____ scabies
          9 ____ *Pneumocystis*
          10 ____ STD

          A. cause of syphilis
          B. removed by surgery
          C. caused by arthropod
          D. sexually transmitted disease
          E. cause of AIDS
          F. anti-HIV drug
          G. a yeast
          H. a protozoan
          I. typical complication in AIDS
             patients
          J. early stage of syphilis

Answers: 1. E   2. F   3. A
                4. B   5. H   6. G
                7. J   8. C   9. I
                10. D

Classification Questions

Answer questions 53- 57 in reference to the five disease listed below:

    a. gonorrhea
    b. syphilis
    c. chlamydia
    d. Herpes simplex
    e. scabies

D   **53.** This is the only one caused by an organism of more than one cell.

M   **54.** This may eventually lead to insanity.

D   **55.** This one remains as a lifelong infection.

M   **56.** This is the major cause of pelvic inflammatory disease.

M   **57.** This is characterized by uneventful symptoms or no symptoms at all.

Answers: 53. e   54. b   55. d
                   56. c   57. a

Selecting the Exception

E   **58.** Three of the four answers listed below are descriptions of bacterial shape. Select the exception.
    a. coccus
    b. bacillus
*   c. pili
    d. spirillum

M   **59.** Four of the five answers listed below are bacterial structures. Select the exception.
    a. cell wall
    b. pilus
    c. flagellum
*   d. capsid
    e. peptidoglycan

E   **60.** Four of the five answers below are usual routes of transmission for HIV. Select the exception.
    a. sharing hypodermic needles
*   b. kissing
    c. sexual intercourse
    d. blood transfer
    e. pregnancy and birth

CHAPTER 17
CELL REPRODUCTION

Multiple-Choice Questions

DIVIDING CELLS: THE BRIDGE BETWEEN GENERATIONS

M   **1.** When a cell undergoes mitosis
- a. the daughter cells have identical genes.
- b. the daughter cell has genes identical to those of the mother cell that produced it.
- c. the amount of cytoplasm in the mother cell and in each daughter cell is equal.
- d. there is an exact duplication and division of all of the organelles between daughter cells.
- * e. the daughter cells have identical genes and these genes are identical to those of the mother cells that produced them.

D   **2.** When a eukaryotic cell divides, the daughter cells
- a. manufacture all the organelles from material in the cytoplasm.
- * b. receive enough of the organelles to start up the new cells and produce additional organelles as needed.
- c. produce individual organelles that attach to the spindle fibers and are distributed just like chromosomes.
- d. produce an equal number of organelles distributed to each cell.
- e. get cellular organelles by an unknown process.

E   **3.** In mitosis, if a parent cell has 16 chromosomes, each daughter cell will have how many chromosomes?
- a. 64
- b. 32
- * c. 16
- d. 8
- e. 4

E   **4.** Chromatids that are attached at the centromere are called what kind of chromatids?
- a. mother
- b. daughter
- * c. sister
- d. programmed
- e. either mother or daughter.

E   **5.** Cells with a double set of genetic information are described by the term
- a. polyploid.
- * b. diploid.
- c. triploid.
- d. haploid.
- e. tetraploid.

D   **6.** Which of these statements concerning the centromere is NOT true?
- a. It appears to join duplicated DNAs.
- * b. It anchors proteins to DNA.
- c. It varies from one type of chromosome to the next.
- d. It is the attachment site for microtubules.
- e. It is temporary.

D   **7.** Which statement is true of homologous chromosomes?
- a. They are of unequal length.
- * b. Each gamete receives one member of each pair.
- c. There are 46 pairs in humans.
- d. They are identical in every way.
- e. They include all chromosomes except X and Y.

E   **8.** Which of the following is NOT a somatic cell?
- a. liver
- b. skin
- c. bone
- d. blood
- * e. sperm

THE CELL CYCLE

M   **9.** DNA replication occurs
- * a. between the gap phases of interphase.
- b. immediately before prophase of mitosis.
- c. during prophase of mitosis.
- d. during prophase of meiosis.
- e. at any time during cell division.

E 10. Chromosomes are duplicated during which period?
   a. M
   b. D
   c. G_1
   d. G_2
 * e. S

M 11. The number of DNA molecules present in a duplicated chromosome is
   a. 1.
   b. undetermined.
   c. half that of an unduplicated chromosome.
 * d. 2.
   e. 4.

A TOUR OF THE STAGES OF MITOSIS

E 12. The spindle apparatus is made of
   a. Golgi bodies.
 * b. microtubules.
   c. endoplasmic reticulum.
   d. nucleoprotein.
   e. chromatids.

E 13. The chromosomes and genes are actually replicated during
   a. anaphase.
   b. metaphase.
 * c. interphase.
   d. prophase.
   e. telophase.

M 14. Which of the following is the proper sequence for mitosis?
   a. metaphase, prophase, anaphase, telophase
   b. metaphase, telophase, prophase, anaphase
 * c. prophase, metaphase, anaphase, telophase
   d. anaphase, metaphase, prophase, telophase
   e. prophase, anaphase, metaphase, telophase

M 15. Mitosis comes from the Greek word *mitos*, which means
   a. divide.
   b. grow.
   c. swell.
 * d. thread.
   e. shrink.

M 16. In eukaryotic cells, which can occur during mitosis?
   a. the duplication of chromatids
   b. the replication of DNA
   c. pairing of homologous chromosomes
 * d. fragmentation and disappearance of nuclear envelope and nucleolus
   e. all of these

E 17. The chromosomes are aligned at the spindle equator during
   a. anaphase.
 * b. metaphase.
   c. interphase.
   d. prophase.
   e. telophase.

E 18. The spindle apparatus becomes visible during
   a. anaphase.
   b. metaphase.
   c. interphase.
 * d. prophase.
   e. telophase.

M 19. The chromosomes detach from one another and become visibly separated during
 * a. anaphase.
   b. metaphase.
   c. interphase.
   d. prophase.
   e. telophase.

E 20. The chromosomes are moving to opposite poles during
 * a. anaphase.
   b. metaphase.
   c. interphase.
   d. prophase.
   e. telophase.

E 21. The chromosomes have arrived at opposite poles during
   a. anaphase.
   b. metaphase.
   c. interphase.
   d. prophase.
 * e. telophase.

E 22. The nuclear membrane reforms during
   a. anaphase.
   b. metaphase.
   c. interphase.
   d. prophase.
 * e. telophase.

M   **23.** Strictly speaking, mitosis and meiosis are divisions of the
   a. nucleus.
   b. cytoplasm.
   c. chromosomes.
*   d. nucleus and chromosomes.
   e. nucleus, chromosomes, and cytoplasm.

D   **24.** In which of the stages below does the chromosome consist of two DNA molecules?
   a. prophase and anaphase
   b. metaphase, prophase, and anaphase
*   c. metaphase and prophase
   d. metaphase, telophase, and prophase
   e. metaphase, telophase, prophase, and anaphase

DIVISION OF THE CYTOPLASM

M   **25.** The distribution of cytoplasm to daughter cells is accomplished during
   a. prokaryotic fission.
   b. mitosis.
   c. meiosis.
*   d. cytokinesis (cytoplasmic division).
   e. karyokinesis.

A CLOSER LOOK AT THE CYTOPLASM

D   **26.** During the "gap" phases of the cell cycle, most of the activity is directed toward
   a. DNA replication.
   b. nuclear membrane synthesis.
   c. resting for the next step.
   d. sorting the chromosomes.
*   e. synthesizing cytoplasmic organelles.

M   **27.** Cell organelles are synthesized during
*   a. the G_1 stage.
   b. the G_2 stage.
   c. the M stage.
   d. the S stage.
   e. all stages.

E   **28.** The proteins that bind tightly to DNA and form "spools" are called
   a. centrioles.
   b. centromeres.
*   c. histones.
   d. motor proteins.
   e. topoisomerases.

M   **29.** The docking sites on the centromeres are called
*   a. kinetochores.
   b. centrioles.
   c. histones.
   d. spindles.
   e. cleavage furrows.

AN OVERVIEW OF MEIOSIS

M   **30.** Which of the following is NOT associated with meiosis?
   a. reduction of number of chromosomes
*   b. somatic cells
   c. sexual reproduction
   d. sperm and egg
   e. germ cells

M   **31.** The essence of meiosis is that
   a. gametes are formed that receive one copy of *each* member of *each* pair of homologous chromosomes.
   b. gametes are formed that are diploid.
   c. each gamete receives one member of *each* pair of homologous chromosomes.
   d. gametes are formed that are haploid.
*   e. each gamete receives one member of *each* pair of homologous chromosomes and gametes are formed that are haploid.

M   **32.** Through meiosis
   a. alternate forms of genes are shuffled.
   b. parental DNA is divided and distributed to forming gametes.
   c. the diploid chromosome number is reduced to haploid.
   d. offspring are provided with new gene combinations.
*   e. all of these are true.

M   **33.** Gamete formation is
   a. the result of the process of mitosis.
   b. the pairing of homologous chromosomes.
*   c. the formation of sex cells.
   d. the fusion of gametes.
   e. a process that occurs only in asexually reproducing forms.

M 34. Sperm are formed directly from the maturation of
    a. sperm mother cells.
*   b. spermatids.
    c. spermatogonial cells.
    d. primary spermatocytes.
    e. secondary spermatocytes.

M 35. The mature ovum is produced by maturation of the
    a. oogonium.
    b. primary oocyte.
    c. secondary polar body.
    d. polar body I.
*   e. none of these

M 36. Gametogenesis is
    a. always the result of the process of meiosis.
    b. the pairing of homologous chromosomes.
*   c. the formation of sex cells.
    d. the fusion of gametes.
    e. a process that occurs only in asexually reproducing forms.

M 37. Which of the following cells is NOT haploid?
    a. secondary spermatocyte
    b. sperm
*   c. primary oocyte
    d. spermatids
    e. polar bodies

E 38. Which of the following will NOT develop into a gamete?
    a. spermatogonium
*   b. polar bodies
    c. oocyte
    d. spermatid
    e. secondary spermatocyte

D 39. Polar bodies
    a. are dumping places for excess genetic material.
    b. have no known biological function.
    c. are produced by meiosis.
    d. will serve as the gametes if something happens to the egg.
*   e. all but "will serve as the gametes if something happens to the egg" are correct.

M 40. Sexual reproduction
    a. leads to uniform characteristics in a population.
*   b. results in new combinations of genetic traits.
    c. produces genetic clones.
    d. requires less tissue differentiation than asexual reproduction.
    e. produces genetic clones and requires less tissue differentiation than asexual reproduction.

M 41. The number of chromosomes found in a eukaryotic cell
    a. indicates the phylogenetic position of the organism.
    b. is constant during the life cycle.
    c. is haploid among asexually reproducing forms and diploid if they reproduce sexually.
*   d. is doubled by fertilization and cut in half by meiosis.
    e. is dependent on the age of the tissue.

D 42. If meiosis did not occur in sexually reproducing organisms,
    a. growth of the zygote would be halted.
    b. mitosis would be sufficient.
    c. gametes would be haploid.
*   d. the chromosome number would double in each generation.
    e. eggs would be haploid, but sperm would be diploid.

A VISUAL TOUR OF THE STAGES OF MEIOSIS

E 43. If a parent cell has 16 chromosomes and undergoes meiosis, the resulting cells will have how many chromosomes?
    a. 64
    b. 32
    c. 16
*   d. 8
    e. 4

KEY EVENTS DURING MEIOSIS I

E 44. Chromatids are
    a. attached at the centriole.
    b. a pair of chromosomes, one from the mother and one from the father.
    c. attached at their centromeres.
    d. identical until crossing over occurs.
*   e. both attached at their centromeres and identical until crossing over occurs.

M   45. Homologous chromosomes
     a. may exchange parts during meiosis.
     b. have alleles for the same characteristics
         even though the gene expression may
         not be the same.
     c. are in pairs, one chromosome of each
         pair from the father and one from the
         mother.
     d. pair up during meiosis.
   * e. all of these

E   46. Copies of chromosomes linked together at
         their centromeres at the beginning of
         meiosis are appropriately called what kind
         of chromatids?
     a. mother
     b. daughter
   * c. sister
     d. homologous
     e. none of these

M   47. Chromosomes of a pair of homologous
         chromosomes may differ from other
         chromosomes in terms of
     a. size.
     b. shape.
     c. alleles they carry.
     d. position of the centromere.
   * e. all of these

D   48. Crossing over
     a. generally results in binary fission.
     b. involves nucleoli.
     c. involves breakages and exchanges being
         made between sister chromatids.
   * d. alters the composition of chromosomes
         and results in new combinations of
         alleles being channeled into the
         daughter cells.
     e. all of these

E   49. Synapsis and crossing over occur during
     a. anaphase I.
     b. metaphase II.
   * c. prophase I.
     d. prophase II.
     e. telophase II.

M   50. Chiasmata provide evidence of
     a. meiosis.
   * b. crossing over.
     c. chromosomal aberration.
     d. fertilization.
     e. spindle fiber formation.

D   51. Crossing over
   * a. increases variability in gametes.
     b. results in only one exchange per
         homologue.
     c. occurs between sister chromatids.
     d. prevents genetic recombination.
     e. is followed immediately by separation
         of each of the chromatids.

E   52. Pairing of homologues and crossing over
         occur during
     a. anaphase I.
     b. metaphase II.
   * c. prophase I.
     d. prophase II.
     e. telophase II.

M   53. Different, or alternative, forms of the same
         gene are called
     a. genetomorphs.
   * b. alleles.
     c. mutants.
     d. chromatids.
     e. homologous.

M   54. Sister chromatids are separated from each
         other during
     a. metaphase I.
     b. anaphase I.
     c. telophase II.
   * d. anaphase II.
     e. metaphase II.

D   55. At the beginning of prophase I, there are
         _____ molecules of DNA in a
         potential human sperm cell.
   * a. 92
     b. 23
     c. 46
     d. half as many (as compared to somatic
         cells)
     e. twice as many (as compared to mature
         sperm)

M   56. Which of the following events does NOT
         occur in prophase II, but does occur in
         prophase I?
     a. crossing over
     b. synapsis
     c. spindle formation
   * d. crossing over and synapsis, only.
     e. crossing over, synapsis, and spindle
         formation..

D 57. Major gene reshuffling takes place during
    *   a   prophase I.
         b.   metaphase I.
         c.   anaphase I.
         d.   metaphase II.
         e.   anaphase II.

M 58. Meiosis typically results in the production of
         a.   2 diploid cells.
         b.   4 diploid cells.
    *   c.   4 haploid cells.
         d.   2 haploid cells.
         e.   1 triploid cell.

E 59. Under favorable conditions, during which phase of meiosis will the chromosomes appear as packets of four chromatids?
         a.   anaphase I
         b.   telophase II
         c.   anaphase II
    *   d.   prophase I
         e.   metaphase II

D 60. Anaphase
         a.   involves the lining up of the chromosomes across the equatorial plate.
         b.   is the same in mitosis and meiosis I and II.
    *   c.   is initiated when the newly divided centromeres begin to move apart.
         d.   results in an unequal distribution of chromosomes to the resulting cells.

D 61. Paired homologous chromosomes are found at the spindle equator during
    *   a.   metaphase I.
         b.   telophase I.
         c.   prophase II.
         d.   metaphase II.
         e.   anaphase II.

E 62. The period that may occur between meiosis I and meiosis II is called
         a.   cytokinesis.
    *   b.   interkinesis.
         c.   synapsis.
         d.   reduction division.
         e.   karyokinesis.

D 63. During meiosis II
         a.   cytokinesis results in the formation of a total of two cells.
    *   b.   sister chromatids of each chromosome are separated from each other.
         c.   homologous chromosomes pair up.
         d.   homologous chromosomes separate.
         e.   sister chromatids exchange parts.

M 64. Which does NOT occur in prophase I of meiosis?
    *   a.   cytokinesis
         b.   formation of groups of four chromatids
         c.   homologue pairing
         d.   crossing over
         e.   condensation of chromatin

M 65. Which is NOT true of human chromosomes?
         a.   The haploid number is 23.
         b.   The diploid number is 46.
         c.   There are 23 pairs of chromosomes.
    *   d.   Human gametes end up with two of each type of 23 chromosomes.
         e.   Human gametes end up one of each type of 23 chromosomes.

D 66. Crossing over is one of the most important events in meiosis because
    *   a.   it produces new arrays of alleles on chromosomes.
         b.   homologous chromosomes must be separated into different daughter cells.
         c.   the number of chromosomes allotted to each daughter cell must be halved.
         d.   homologous chromatids must be separated into different daughter cells.
         e.   all of these

M 67. Which does NOT produce variation?
         a.   crossing over
         b.   random alignment of chromosomes during meiosis
    *   c.   mitosis
         d.   genetic recombination of alleles
         e.   sexual reproduction

M 68. Maternal and paternal chromosomes are shuffled most during
         a.   anaphase II.
    *   b.   metaphase I.
         c.   prophase I.
         d.   telophase II.
         e.   interphase.

MEIOSIS AND MITOSIS COMPARED

D   **69.** In comparing mitosis and meiosis, which of the following statements is true?
     a. Meiosis I is more like mitosis than is meiosis II.
     b. Both processes result in four cells.
     c. Synapsis occurs in both.
     d. Chromatids are present only in mitosis.
     *   e. Meiosis II resembles mitosis.

Matching Questions

E   **70.** Match each numbered item in the left-hand column with a lettered item from the right-hand column.

1 _____ cell cycle
2 _____ centromere
3 _____ chromatid
4 _____ cytokinesis
5 _____ metaphase
6 _____ microtubules
7 _____ prophase
8 _____ telophase
9 _____ anaphase

     A. cytoplasm apportioned between the two daughter cells
     B. final phase of mitosis; daughter nuclei re-form
     C. two sister chromatids are joined here
     D. chromosomes condense and mitotic spindle begins to form
     E. chromosomes line up at spindle equator
     F. sister chromatids separate; move to opposite spindle poles now
     G. about 25 nm in diameter; form mitotic spindle
     H. half of a chromosome in prophase
     I. mitosis plus interphase

Answers: 1. I   2. C   3. H
              4. A   5. E   6. G
              7. D   8. B   9. F

Classification Questions

Answer questions 71-80 in reference to the eukaryotic cell cycle. Each question has only one BEST answer.

     a. G_2
     b. mitosis
     c. S
     d. G_1
     e. cytokinesis

E   **71.** Period when DNA is duplicated.

E   **72.** Period when interphase ends in the parent cell.

M   **73.** Event that forms two daughter cytoplasmic masses.

M   **74.** Period of cell growth before DNA duplication.

M   **75.** Period after DNA is duplicated.

E   **76.** Period of nuclear division.

M   **77.** Period when interphase begins in a daughter cell.

E   **78.** Period commonly followed by cytokinesis.

E   **79.** The period in which metaphase occurs.

D   **80.** The period prior to mitosis.

Answers: 71. c   72. a   73. e
              74. d   75. a   76. b
              77. d   78. b   79. b
              80. a

The stages of mitosis plus interphase are listed under
a–e below. Answer questions 81-91 with reference to these
phases.

    a. interphase
    b. prophase
    c. metaphase
    d. anaphase
    e. telophase

E   **81.** During this stage homologous pairs of
         chromosomes are lined up on the equatorial
         plate.

M   **82.** Chromosomes replicate during this phase.

M   **83.** Genes replicate during this phase.

E   **84.** DNA replicates during this phase.

E   **85.** Condensation and shortening of
         chromosomes occurs during this phase.

E   **86.** Spindle fibers first appear during this stage.

M   **87.** During this phase the centromeres break
         apart as the separated sister chromatids
         begin to move to opposite poles.

E   **88.** The microtubular spindle develops during
         this phase.

E   **89.** Sister chromatids joined at their centromeres
         are attached to spindle fibers during this
         phase.

M   **90.** Cytokinesis occurs as this phase of mitosis
         proceeds.

E   **91.** New daughter nuclear membranes form
         during this phase.

Answers: 81. c   82. a   83. a
              84. a   85. b   86. b
              87. d   88. b   89. c
              90. e   91. e

Referring to mammalian reproduction, answer 92-94 by
using the five items listed below.

    I.   sperm
    II. mature ova
    III. primary oocytes
    IV. primary spermatocytes
    V.   zygotes

E   **92.** During fertilization which two items
         combine to form a fertilized egg?
      * a. I and II
         b. I and III
         c. I and IV
         d. II and IV
         e. III and IV

M   **93.** Which item or items are the same as a
         fertilized egg?
         a. II only
         b. III only
      * c. V only
         d. II and III
         e. III and V

D   **94.** Which is a normal sequence of
         development?
         a. I >>> II >>> III
         b. I >>> IV >>> V
         c. II >>> III >>> V
      * d. III >>> II and IV >>> I
         e. I >>> IV and II >>> V

Answer questions 95-98 by using the five numbers below.

    a. 10
    b. 20
    c. 40
    d. 60
    e. 80

M   **95.** How many sperm would eventually be
         produced from 20 spermatids?

M   **96.** How many sperm would eventually be
         produced from 20 primary spermatocytes?

M   **97.** How many ova (eggs) would eventually
         result from 20 secondary oocytes?

M   **98.** How many ova (eggs) would eventually
         result from 20 primary oocytes?

Answers: 95. b   96. e   97. b
              98. b

Some of the stages of meiosis are listed under a–e below. Answer questions 99-106 with reference to these phases of meiosis.

    a. prophase I
    b. prophase II
    c. anaphase II
    d. anaphase I
    e. telophase I

E   **99.** The formation of groups of four chromatids occurs during this stage.

E   **100.** Recombination via crossing over occurs during this stage.

D   **101.** By the end of this phase the number of homologous chromosomes is reduced in half.

M   **102.** During this stage the sister chromatids separate.

D   **103.** Following this phase, each individual *cell* is haploid.

M   **104.** Chiasmata are present during this stage.

D   **105.** During this phase the centromeres separate.

D   **106.** New genetic combinations, upon which natural selection can act, are present after this stage.

Answers:
99.	a	100.	a	101.	d
102.	c	103.	e	104.	a
105.	c	106.	a		

Selecting the Exception

E   **107.** Four of the five answers listed below are stages of actual nuclear division. Select the exception.
    a. anaphase
    b. prophase
 * c. interphase
    d. telophase
    e. metaphase

D   **108.** Four of the five answers listed below are related by a common phase of mitosis. Select the exception.
    a. microtubules start to assemble outside the nucleus
 * b. division of centromere
    c. disappearance of nucleolus
    d. disappearance of nuclear membrane
    e. shortening and condensation of a visible chromosome

M   **109.** Four of the five answers listed below are periods of the same cycle. Select the exception.
    a. G_1
    b. M
 * c. R
    d. S
    e. G_2

D   **110.** Four of the five answers listed below are related by a common phase of mitosis. Select the exception.
    a. chromosomes decondense
    b. spindle microtubules disappear
    c. nucleolus reappears
 * d. chromosomes separate
    e. nuclear envelope reforms

M   **111.** Four of the five answers listed below are events occurring during mitosis. Select the exception.
 * a. chromosome replication
    b. division of centromere
    c. lining chromosomes up at the cellular equator
    d. spindle microtubules attach to centromeres
    e. chromosomes migrate to opposite ends of the cell

D   **112.** Four of the five answers listed below assist in chromosome movement. Select the exception.
    a. microtubule
    b. spindle microtubules
    c. centromeres
    d. centriole
 * e. nuclear envelope

D 113. Four of the five answers listed below are related by a common phase of mitosis. Select the exception.
    *   a. chromosomes align at the spindle equator
        b. sister chromatids become individual chromosomes
        c. centromeres divide
        d. the chromosomes move apart
        e. spindle microtubules shorten, pulling chromosomes toward the poles

D 114. Four of the five answers listed below are related by a common division association. Select the exception.
        a. mitochondria
    *   b. chromosomes
        c. ribosomes
        d. plastids
        e. microbodies

M 115. Three of the four answers listed below concern cells with two chromosome sets. Select the exception.
        a. zygote
        b. somatic cells
    *   c. gamete
        d. diploid

D 116. Four of the five answers listed below are characteristic of meiosis. Select the exception.
        a. involves two divisions
        b. reduces the number of chromosomes
    *   c. results in producing genetically identical cells
        d. produces haploid cells
        e. occurs in the gonads

M 117. Four of the five answers listed below are terms describing haploid cells. Select the exception.
        a. ovum
    *   b. primary spermatocyte
        c. spermatid
        d. polar body
        e. secondary spermatocyte

M 118. Four of the five answers listed below are related to pairing of chromosomes. Select the exception.
        a. synapsis
        b. crossing over
        c. exchange of genes
        d. pairing of homologues
    *   e. interkinesis

D 119. Four of the five answers listed below are related to the process of synapsis. Select the exception.
        a. genetic recombination
        b. increase in variability
        c. exchange of genes
    *   d. identical daughter cells
        e. chiasmata

CHAPTER 18

OBSERVABLE PATTERNS OF INHERITANCE

Multiple-Choice Questions

PATTERNS OF INHERITANCE

E **1.** A locus is
    a. a recessive gene.
    b. an unmatched allele.
    c. a sex chromosome.
  * d. the location of an allele on a chromosome.
    e. a dominant gene.

M **2.** Mendel's study of genetics differed from those of his contemporaries because he
    a. used only pure-breeding parents.
    b. examined several different traits at the same time.
  * c. kept careful records and analyzed the data statistically.
    d. worked on plants rather than animals.
    e. confirmed the blending theory of inheritance.

E **3.** Which of the following descriptions of Mendel is incorrect?
  * a. He was simply lucky to work out the laws of genetics.
    b. He focused on contrasting phenotypic characteristics.
    c. He demonstrated that the blending theory of inheritance was wrong.
    d. He kept exact mathematical data and was the first scientist to utilize numerical analysis of results.
    e. He was a monk, a science teacher, and a gardener.

E **4.** Which organism did Mendel use to work out the laws of segregation and independent assortment?
    a. the fruit fly
    b. *Neurospora*
  * c. the garden pea
    d. the chicken
    e. *E. coli*

M **5.** The pea plant was an excellent choice for Mendel's experiments because
    a. true-breeding varieties were available.
    b. the plant can self-fertilize.
    c. it can be cross-fertilized.
    d. true-breeding varieties were available and it can be cross-fertilized.
  * e. true-breeding varieties were available, and the plant can be both cross-fertilized and self-fertilized.

E **6.** Various forms of a gene at a given locus are called
    a. chiasmata.
  * b. alleles.
    c. autosomes.
    d. loci.
    e. chromatids.

M **7.** Diploid organisms
    a. have corresponding alleles on homologous chromosomes.
    b. are usually the result of the fusion of two haploid gametes.
    c. have two sets of chromosomes.
    d. have pairs of homologous chromosomes.
  * e. all of these

E **8.** Which of the following genotypes is homozygous?
    a. *AaBB*
    b. *aABB*
  * c. *aaBB*
    d. *aaBb*
    e. *AaBb*

M **9.** The most accurate description of an organism with genotype *AaBb* is
    a. homozygous dominant.
  * b. heterozygous.
    c. heterozygous dominant.
    d. homozygous recessive.
    e. heterozygous recessive.

D **10.** Gene *A* occurs on chromosome #5, gene *B* is on chromosome #21. Therefore, these two parts of the chromosomes CANNOT be
    a. genes.
    b. dominant.
    c. loci.
  * d. alleles.
    e. recessive.

E   11. If short hair (L) is dominant to long hair (l), animals LL and Ll have the same
    a. parents.
    b. genotypes.
 *   c. phenotypes.
    d. alleles.
    e. genes.

E   12. An individual with a genetic makeup of aa BB is said to be
 *   a. pure-breeding.
    b. recessive.
    c. hybrid.
    d. dihybrid.
    e. heterozygous.

MENDEL'S THEORY OF SEGREGATION

M   13. If R is dominant to r, the offspring of the cross of RR with rr will
    a. be homozygous.
 *   b. display the same phenotype as the RR parent.
    c. display the same phenotype as the rr parent.
    d. have the same genotype as the RR parent.
    e. have the same genotype as the rr parent.

M   14. Mendel found that pea plants expressing a recessive trait
 *   a. were pure-breeding.
    b. appeared only in the first generation of a cross between two pure-breeding plants expressing contrasting forms of a trait.
    c. disappeared after the second generation.
    d. could be produced only if one of the parents expressed the recessive trait.
    e. none of these

M   15. If tall (D) is dominant to dwarf (d), and two homozygous varieties DD and dd are crossed, then what kind of offspring will be produced?
    a. all intermediate forms
 *   b. all tall
    c. all dwarf
    d. 1/2 tall, 1/2 dwarf
    e. 3/4 tall, 1/4 dwarf

M   16. If all offspring of a cross have the genotype Aa, the parents of the crosses would most likely be
 *   a. AA x aa.
    b. Aa x Aa.
    c. Aa x aa.
    d. AA x Aa.
    e. none of these

D   17. The principle of segregation applies most specifically to events occurring in preparation of
    a. offspring.
    b. zygotes.
    c. homologous chromosomes.
 *   d. gametes.
    e. loci.

D   18. If Mendel had not examined the _____ generation, he would not have discovered his principle of segregation.
    a. P_1
    b. H_1
    c. A_1
    d. F_1
 *   e. F_2

E   19. The theory of segregation
    a. deals with the alleles governing two different traits.
    b. applies only to linked genes.
    c. applies only to sex-linked genes.
 *   d. explains the behavior of a pair of alleles during meiosis.
    e. none of these

DOING GENETIC CROSSES AND FIGURING PROBABILITIES

E   20. Hybrid organisms produced from a cross between two pure-breeding organisms belong to which generation?
    a. P_1
    b. H_1
    c. A_1
 *   d. F_1
    e. F_2

M   21. According to Mendel, what kind of genes "disappear" in F_1 pea plants?
    a. sex-linked
    b. dominant
 *   c. recessive
    d. codominant
    e. lethal

E   **22.** The F$_2$ phenotypic ratio of a monohybrid cross is
    a. 1:1.
    b. 2:1.
    c. 9:3:3:1.
    d. 1:2:1.
    * e. 3:1.

M   **23.** For Mendel's explanation of inheritance to be correct,
    a. the genes for the traits he studied had to be located on the same chromosome.
    * b. which gametes combine at fertilization had to be due to chance.
    c. genes could not be transmitted independently of each other.
    d. only diploid organisms would demonstrate inheritance patterns.
    e. none of these

M   **24.** In a Punnett square, the letters within the little boxes represent
    * a. offspring genotypes.
    b. parental genotypes.
    c. gametes.
    d. offspring phenotypes.
    e. parental phenotypes.

E   **25.** If short hair *(L)* is dominant to long hair *(l)*, then what fraction of the offspring produced by a cross of *Ll* x *ll* will be homozygous dominant?
    a. 1/2
    b. 1/4
    c. 1/3
    * d. none (no chance of this offspring)
    e. none of these is correct

M   **26.** If short hair *(L)* is dominant to long hair *(l)*, then to determine the genotype of a short-haired animal it should be crossed with
    a. *LL.*
    b. *Ll.*
    * c. *ll.*
    d. all of these
    e. none of these

E   **27.** What fraction of the time will the cross of *Aa Bb Cc* with *Aa Bb Cc* produce an offspring of genotype *aa bb cc* ?
    * a. 1/64
    b. 1/32
    c. 3/64
    d. 1/16
    e. 9/64

M   **28.** What fraction of the time will the cross of *Aa Bb Cc* with *Aa Bb Cc* produce an offspring of genotype *Aa bb CC* ?
    a. 1/64
    * b. 1/32
    c. 3/64
    d. 1/16
    e. 9/64

D   **29.** What fraction of the time will the cross of *Aa Bb Cc* with *Aa Bb Cc* produce an offspring that expresses the dominant traits *A* and *B* and *cc* (*A_ B_ cc*)?
    a. 1/32
    b. 3/64
    c. 1/16
    * d. 9/64
    e. 27/64

D   **30.** What fraction of the time will the cross of *Aa Bb Cc* with *Aa Bb Cc* produce an offspring that expresses the phenotype represented by the dominant gene *C* (*aa bb C_*)?
    a. 1/32
    * b. 3/64
    c. 1/16
    d. 9/64
    e. 27/64

D   **31.** What fraction of the time will the cross of *Aa Bb Cc* with *Aa Bb Cc* produce an offspring that expresses all three dominant genes?
    a. 3/64
    b. 1/16
    c. 1/8
    d. 9/64
    * e. 27/64

D   **32.** What fraction of the time will the cross of *Aa Bb Cc* with *Aa Bb Cc* produce an offspring that is pure-breeding?
    a. 3/64
    b. 1/16
    * c. 1/8
    d. 9/64
    e. 27/64

D   **33.** The chance of producing an offspring of genotype *Aa BB cc* from a cross of *Aa Bb Cc* with *Aa Bb Cc* is
    a. 1/64.
    * b. 1/32.
    c. 3/64.
    d. 1/16.
    e. 3/32.

D **34.** The chance of producing an offspring of genotype *Aa Bb cc* from a cross of *Aa BB Cc* with *Aa BB Cc* is
    a. 1/32.
    b. 1/16.
    c. 3/32.
    d. 1/8.
*   e. none (no chance of this offspring)

D **35.** What fraction of the time will a cross of *Aa Bb Cc* with *Aa BB cc* produce an offspring of genotype *Aa Bb Cc*?
    a. 1/32
    b. 1/16
    c. 3/32
*   d. 1/8
    e. none (no chance of this offspring)

D **36.** What fraction of the time will a cross of *Aa BB cc* with *Aa Bb CC* produce an offspring of genotype *Aa Bb CC*?
    a. 1/32
    b. 1/16
    c. 3/32
    d. 1/8
*   e. none (no chance of this offspring)

THE TESTCROSS: A TOOL FOR DISCOVERING GENOTYPES

D **37.** Short hair *(L)* is dominant to long hair *(l)*. If a short-haired animal of unknown origin is crossed with a long-haired animal and they produce one long-haired and one short-haired offspring, this would indicate that
    a. the short-haired animal was pure-breeding.
*   b. the short-haired animal was not pure-breeding.
    c. the long-haired animal was not pure-breeding.
    d. the long-haired animal was pure-breeding.
    e. none of these can be determined with two offspring

M **38.** The results of a testcross reveal that all offspring resemble the parent being tested. That parent necessarily is
    a. heterozygous.
    b. polygenic.
*   c. homozygous.
    d. recessive.

D **39.** A testcross involves
    a. two F_1 hybrids.
    b. an F_1 hybrid and an F_2 offspring.
    c. two parental organisms.
    d. an F_1 hybrid and the homozygous dominant parent.
*   e. an F_1 hybrid and an organism that is homozygous recessive for that trait.

D **40.** A testcross consists of
    a. a cross of two pure-breeding forms to find out which form of a gene is dominant.
    b. a cross between two unknown forms to determine their genotypes.
    c. a cross between an offspring and its parent.
*   d. a cross of an F_1 hybrid to an individual that is homologous recessive.
    e. a cross of two F_2 individuals to produce an F_3 generation.

M **41.** For monohybrid experiments, a testcross could result in which of the following ratios?
*   a. 1:1
    b. 2:1
    c. 9:3:3:1
    d. 1:2:1
    e. 3:1

M **42.** If all the offspring of a testcross are alike and resemble the organism being tested, then that parent is
*   a. homozygous dominant.
    b. homozygous recessive.
    c. heterozygous.
    d. recessive.
    e. incompletely dominant.

M **43.** Assume short hair *(L)* is dominant to long hair *(l)* and black hair *(B)* is dominant to brown *(b)*. If you found a black short-haired animal, you could determine its genotype by crossing it to an animal with a genotype of
    a. *LL BB*.
    b. *ll BB*.
    c. *ll Bb*.
*   d. *ll bb*.
    e. *LL bb*.

M   **44.** If all the offspring of a cross had the genotype *Aa Bb*, the parents of the cross would most likely be
   a. *AA BB* x *aa bb*.
   b. *AA bb* x *aa BB*.
   c. *Aa Bb* x *Aa Bb*.
   d. *Aa bb* x *aa Bb*.
   * e. both a or b, but not c or d

INDEPENDENT ASSORTMENT

D   **45.** Some dogs have erect ears; others have drooping ears. Some dogs bark when following a scent; others are silent. Erect ears and barking are due to dominant alleles located on different chromosomes. A dog homozygous for both dominant traits is mated to a droopy-eared, silent follower. The phenotypic ratio expected in the F_1 generation is
   a. 9:3:3:1.
   * b. 100 percent of one phenotype.
   c. 1:1.
   d. 1:2:1.
   e. none of these

D   **46.** Some dogs have erect ears; others have drooping ears. Some dogs bark when following a scent; others are silent. Erect ears and barking are due to dominant alleles located on different chromosomes. If two dihybrids are crossed
   a. the most common phenotype is drooping ears and barking.
   * b. all droopy-eared silent dogs are pure-breeding.
   c. the least common phenotype is drooping ears and barking.
   d. there will be no phenotypes or genotypes that resemble the original parents.
   e. there will be no offspring that resemble the F_1 generation.

M   **47.** Mendel's principle of independent assortment states that
   a. one allele is always dominant to another.
   b. hereditary units from the male and female parents are blended in the offspring.
   c. the two hereditary units that influence a certain trait segregate during gamete formation.
   * d. each hereditary unit is inherited separately from other hereditary units.

D   **48.** Which of the following would be an exception to the principle of independent assortment?
   a. dominance
   b. recessiveness
   c. incomplete dominance
   d. pleiotropy
   * e. linkage

D   **49.** Mendel's dihybrid crosses provided indirect evidence for all but which one of the following?
   a. independent assortment
   b. dominance
   * c. linkage
   d. presence of two factors in parents and offspring
   e. segregation of factors

D   **50.** In cocker spaniels, black coat color *(B)* is dominant over red *(b)*, and solid color *(S)* is dominant over spotted *(s)*. If a red male was crossed with a black female to produce a red spotted puppy, the genotypes of the parents (with male genotype first) would be
   a. *Bb Ss* x *Bb Ss*.
   * b. *bb Ss* x *Bb Ss*.
   c. *bb ss* x *Bb Ss*.
   d. *bb Ss* x *Bb ss*.
   e. *Bb ss* x *Bb ss*.

D   **51.** In cocker spaniels, black coat color *(B)* is dominant over red *(b)*, and solid color *(S)* is dominant over spotted *(s)*. If a red spotted male was crossed with a black solid female and all the offspring from several crosses expressed only the dominant traits, the genotype of the female would be
   * a. *BB S S*.
   b. *Bb S S*.
   c. *Bb S s*.
   d. *BB S s*.
   e. none of these

D   **52.** In cocker spaniels, black coat color *(B)* is dominant over red *(b)*, and solid color *(S)* is dominant over spotted *(s)*. If two black solid dogs were crossed several times and the total offspring were eighteen black solid and five black spotted puppies, the genotypes of the parents would most likely be
   a. *Bb Ss* x *Bb Ss*.
   b. *Bb Ss* x *Bb S S*.
   c. *BB Ss* x *Bb ss*.
   * d. *BB Ss* x *Bb Ss*.
   e. *Bb ss* x *Bb S S*.

M 53. In cocker spaniels, black coat color (B) is dominant over red (b), and solid color (S) is dominant over spotted (s). If two dihybrids (Bb Ss) were crossed, the most common phenotype would be
* a. black and solid.
   b. black and spotted.
   c. red and solid.
   d. red and spotted.
   e. none of these

M 54. In cocker spaniels, black coat color (B) is dominant over red (b), and solid color (S) is dominant over spotted (s). If two dihybrids (Bb Ss) were crossed, which would be produced?
   a. black and spotted pure-breeding forms
   b. black and solid pure-breeding forms
   c. red and solid pure-breeding forms
   d. red and spotted pure-breeding forms
* e. all of these

D 55. In cocker spaniels, black coat color (B) is dominant over red (b), and solid color (S) is dominant over spotted (s). If two dihybrids (Bb Ss) were crossed, what fraction of the black solid offspring would be homozygous?
   a. 4/16
   b. 9/16
* c. 1/9
   d. 3/16
   e. 3/4

D 56. In cocker spaniels, black coat color (B) is dominant over red (b), and solid color (S) is dominant over spotted (s). In the F_2 generation of a cross between BB ss with bb S S, what fraction of the offspring would be expected to be black and spotted?
   a. 1/16
   b. 9/16
   c. 1/9
* d. 3/16
   e. 3/4

M 57. In cocker spaniels, black coat color (B) is dominant over red (b), and solid color (S) is dominant over spotted (s). A cross of Bb Ss with bb ss would produce the phenotypic ratio
   a. 9:3:3:1.
* b. 1:1:1:1.
   c. 1:2:1.
   d. 3:1.
   e. none of these

D 58. In cocker spaniels, black coat color (B) is dominant over red (b), and solid color (S) is dominant over spotted (s). If Bb Ss were crossed with Bb ss, the chance that a black solid individual would be produced is
   a. 3/16.
   b. 1/3.
   c. 9/16.
* d. 3/8.
   e. 1/16.

M 59. The theory of independent assortment
* a. cannot be demonstrated in a monohybrid cross.
   b. is illustrated by the behavior of linked genes.
   c. indicates that the expression of one gene is independent of the action of another gene.
   d. states that alleles for the same characteristic separate during meiosis.
   e. is negated by epistasis.

M 60. Mendel's dihybrid crosses, but not his monohybrid crosses, showed that
   a. some genes were linked together.
   b. the two alleles controlling a trait were divided equally among the gametes.
* c. alleles for different traits were inherited independently.
   d. one of the pair of alleles is dominant to the other.
   e. the crossing of two different homozygous forms will not produce any offspring in the first generation that will look like either of the parents.

A CLOSER LOOK AT INDEPENDENT ASSORTMENT

D 61. In the second generation of a cross of DD RR with dd rr, the most common genotype would be
   a. DD RR.
   b. Dd RR.
* c. Dd Rr.
   d. dd RR.
   e. dd Rr.

E 62. The usual F_2 phenotypic ratio of a dihybrid cross is
   a. 1:1.
   b. 2:1.
* c. 9:3:3:1.
   d. 1:2:1.
   e. 3:1.

D   **63.** Individuals with the genotype *Gg Hh Ii Jj* will produce how many different kinds of gametes?
  - a. 2
  - b. 4
  - c. 6
  - d. 8
  - *   e. 16

M   **64.** An individual with a genotype of *Aa Bb CC* is able to produce how many different kinds of gametes?
  - a. 2
  - b. 3
  - *   c. 4
  - d. 7
  - e. 8

M   **65.** A dihybrid cross of two contrasting pure-breeding organisms
  - a. produces homozygous offspring.
  - b. must produce a phenotype different from either pure-breeding parent.
  - *   c. results in the disappearance of the recessive traits for the first generation.
  - d. takes place only in the laboratory under precisely controlled conditions.
  - e. will result in the immediate formation of another pure-breeding variety.

MULTIPLE EFFECTS OF SINGLE GENES

D   **66.** Coat color in one breed of mice is controlled by incompletely dominant alleles so that yellow and white are homozygous, while cream is heterozygous. The cross of two cream individuals will produce
  - a. all cream offspring.
  - b. equal numbers of white and yellow mice, but no cream.
  - c. equal numbers of white and cream mice.
  - d. equal numbers of yellow and cream mice.
  - *   e. equal numbers of white and yellow mice, with twice as many creams as the other two colors.

D   **67.** An incompletely dominant gene controls the color of chickens so that *BB* produces black, *Bb* produces a slate-gray color called blue, and *bb* produces splashed white. A second gene controls comb shape, with the dominant gene *R* producing a rose comb and *r* producing a single comb. If a pure-breeding black chicken with a rose comb is mated to a splashed white chicken with a single comb in the F_2 generation, what fraction of the offspring will be black with rose comb?
  - a. 9/16
  - b. 3/8
  - *   c. 3/16
  - d. 1/8
  - e. 1/16

D   **68.** An incompletely dominant gene controls the color of chickens so that *BB* produces black, *Bb* produces a slate-gray color called blue, and *bb* produces splashed white. A second gene controls comb shape, with the dominant gene *R* producing a rose comb and *r* producing a single comb. If a pure-breeding black chicken with rose comb is mated to a splashed white chicken with a single comb in the F_2 generation, what fraction of the offspring will be black with single comb?
  - a. 9/16
  - b. 3/8
  - c. 3/16
  - d. 1/8
  - *   e. 1/16

D   **69.** An incompletely dominant gene controls the color of chickens so that *BB* produces black, *Bb* produces a slate-gray color called blue, and *bb* produces splashed white. A second gene controls comb shape, with the dominant gene *R* producing a rose comb and *r* producing a single comb. If a pure-breeding black chicken with a rose comb is mated to a splashed white chicken with a single comb in the F_2 generation, what fraction of the offspring will be blue with single comb?
  - a. 9/16
  - b. 3/8
  - c. 3/16
  - *   d. 1/8
  - e. 1/16

D   70. An incompletely dominant gene controls
         the color of chickens so that *BB* produces
         black, *Bb* produces a slate-gray color called
         blue, and *bb* produces splashed white. A
         second gene controls comb shape, with the
         dominant gene *R* producing a rose comb and
         r producing a single comb. If a pure-
         breeding black chicken with a rose comb is
         mated to a splashed white chicken with a
         single comb in the F_2 generation, what
         fraction of the offspring will be blue with
         rose comb?
         a. 9/16
     *   b. 3/8
         c. 3/16
         d. 1/8
         e. 1/16

D   71. Susan, a mother with type B blood, has a
         child with type O blood. She claims that
         Craig, who has type A blood, is the father.
         He claims that he cannot possibly be the
         father. Further blood tests ordered by the
         judge reveal that Craig is AA. The judge
         rules that
         a. Susan is right and Craig must pay child
             support.
     *   b. Craig is right and doesn't have to pay
             child support.
         c. Susan cannot be the real mother of the
             child; there must have been an error
             made at the hospital.
         d. it is impossible to reach a decision
             based on the limited data available.
         e. none of these

M   72. If a child has an AB blood type, the parents
         a. must both have different blood types.
         b. must be A and B, but not AB.
         c. must both be AB.
         d. can be any blood type.
     *   e. can have different blood types, but
             neither can be blood type O.

E   73. Blood types (A, B, and O) are controlled by
         a. sex-linked genes.
         b. linked genes.
         c. multiple genes.
     *   d. multiple alleles.
         e. none of these

M   74. The ABO blood types are controlled by
         a. single genes.
         b. multiple alleles.
         c. incomplete dominance.
         d. codominance.
     *   e. both multiple alleles and codominance.

D   75. The ABO blood types have _____
         different genotypes.
         a. 4
     *   b. 6
         c. 8
         d. 12
         e. 16

M   76. If a child belonged to blood type O, he or
         she could NOT have been produced by
         which set of parents?
         a. Type A mother and type B father
         b. Type A mother and type O father
     *   c. Type AB mother and type O father
         d. Type O mother and type O father
         e. a and c could not, but both b and d
             could produce a type O child

D   77. The number of different alleles for ABO
         blood types in the total human population
         is
         a. 4.
         b. 6.
         c. 9.
         d. undetermined.
     *   e. 3.

D   78. If a woman of blood type A has a child of
         blood type O, the father may belong to
         blood type
         a. A, AB, O, but not B.
         b. O only.
     *   c. A, B, O, but not AB.
         d. any blood type other than type A

D   79. If one parent has type A blood and the other
         parent has type B, then which of the
         following is possible in the children?
         a. only AB
         b. A and AB
         c. B and AB
     *   d. A, B, AB, O
         e. only O

M   80. If a child belonged to blood type O, he or
         she could not have been produced by which
         set of parents?
         a. Type A mother and type B father
         b. Type A mother and type O father
     *   c. Type AB mother and type O father
         d. Type O mother and type O father
         e. a and c could not, but both b and d
             could produce a type O child

E   81. A gene that produces multiple effects is
         called
         a. a multiple allele.
         b. an autosome.
         c. an epistatic gene.
      * d. a pleiotropic gene.
         e. an incompletely dominant gene.

E   82. Multiple effects of a single gene is known
         as
         a. expressivity.
         b. penetrance.
         c. codominance.
      * d. pleiotropy.
         e. multiple alleles.

E   83. Pleiotropic genes
         a. act on secondary sexual characteristics.
      * b. influence more than one aspect of
            phenotype.
         c. are additive.
         d. produce lethal effects when
            homozygous.
         e. none of these

HOW CAN WE EXPLAIN LESS PREDICTABLE VARIATIONS?

E   84. A bell-shaped curve of phenotypic variation
         is a representation of
         a. incomplete dominance.
      * b. continuous variation.
         c. multiple alleles.
         d. epistasis.
         e. environmental variables on phenotypes.

E   85. The percentage of individuals in which a
         particular genotype is expressed is
         a. pleiotropy.
         b. genomic imprinting.
         c. codominance.
      * d. penetrance.
         e. linkage.

M   86. Persons who inherit a gene for extra fingers
         or toes __?__ have the physical
         manifestation.
      * a. may or may not
         b. will always
         c. cannot
         d. hardly ever

E   87. What is the name given to the phenomenon
         in which it appears that the effect of a gene
         in an offspring depends on which parent
         contributed the gene?
         a. campodactyly
      * b. genomic imprinting
         c. variable expressivity
         d. pleiotropy
         e. codominance

E   88. Traits which are controlled by more than
         one gene are called
         a. codominant.
         b. pleiotropic.
      * c. polygenic.
         d. penetrant.
         e. independent.

E   89. Human skin and eye color are traits that are
         a. the result of polygenic inheritance.
         b. examples of continuous variation.
         c. incompletely penetrant.
      * d. the result of polygenic inheritance and
            examples of continuous variation.
         e. the result of polygenic inheritance,
            examples of continuous variation, and
            incompletely penetrant.

E   90. Which of the following statements
         concerning genes and behavior is true?
         a. Most genes for desirable behavior are
            dominant.
      * b. The relationship between genes and
            behavior is not firmly established.
         c. Studies of twins prove that behavior is
            genetically programmed.
         d. Because behavior genes are recessive,
            experiments should be easy to design.
         e. all of these are correct

Problems

M   91. In a certain plant, when individuals with
         blue flowers are crossed with individuals
         with blue flowers, only blue flowers are
         produced. Plants with red flowers crossed
         with plants with red flowers sometimes
         produce only red flowers, while other times
         they produce either red or blue flowers.
         When plants with red flowers are crossed
         with plants with blue flowers, sometimes
         only red flowers are produced; other times
         either red or blue flowers are produced.
         Which gene is dominant?

M   **92.** Which is easier to establish in a pure-breeding population, a dominant or a recessive gene?

M   **93.** Tall (*D*) is dominant to dwarf (*d*). Give the F_2 genotypic and phenotypic ratios of a cross between a pure-breeding tall plant and a pure-breeding dwarf plant.

M   **94.** If wire hair (*W*) is dominant to smooth hair (*w*) and you find a wire-haired puppy, how would you determine its genotype by a genetic breeding experiment? Give both the genotype and phenotype involved with the cross with the unknown.

M   **95.** In poultry, rose comb is controlled by a dominant allele and its recessive allele controls single comb.
     (a)   Give the genotype and phenotype produced from crossing a pure-breeding rose comb chicken with a pure-breeding single comb chicken.
     (b)   Give the results of the backcross of the F_1 hybrid with both pure-breeding parents.

M   **96.** If black fur color is controlled by a dominant allele (*B*) and brown by its recessive allele (*b*), give the genotypes of the parents and offspring of a cross of a black male with a brown female that produces 1/2 black offspring and 1/2 brown offspring.

D   **97.** If 2 spot (*S*) is dominant to 4 spot (*s*), give the genotypes for the parents in the following crosses:
     (a)   2 spot x 2 spot yields 2 spot and 4 spot
     (b)   2 spot x 4 spot yields only 2 spot
     (c)   2 spot x 4 spot yields 2 spot and 4 spot
     (d)   2 spot x 2 spot yields only 2 spot
     (e)   4 spot x 4 spot yields only 4 spot

D   **98.** In humans, normal skin pigmentation is influenced by a dominant gene (*C*), which allows pigmentation to develop. All individuals who are homozygous for the recessive allele (*c*) are unable to produce an enzyme needed for melanin formation and are therefore referred to as albino. Two normal parents produce an albino child. What are the chances that the next child will be an albino?

D   **99.** The allele for albinism (*c*) is recessive to the allele for normal pigmentation (*C*). A normally pigmented woman whose father is an albino marries an albino man whose parents are normal. They have three children, two normal and one albino. Give the genotypes for each person listed.

D   **100.** In garden peas, one pair of alleles controls the height of the plant and a second pair of alleles controls flower color. The allele for tall (*D*) is dominant to the allele for dwarf (*d*), and the allele for purple (*P*) is dominant to the allele for white (*p*). A tall plant with purple flowers crossed with a dwarf plant with white flowers produces 1/2 tall with purple flowers and 1/2 tall with white flowers. What is the genotype of the parents?

D   **101.** In garden peas, one pair of alleles controls the height of the plant and a second pair of alleles controls flower color. The allele for tall (*D*) is dominant to the allele for dwarf (*d*), and the allele for purple (*P*) is dominant to the allele for white (*p*). A tall plant with white flowers crossed with a dwarf plant with purple flowers produces all tall offspring with purple flowers. What is the genotype of the parents?

D   **102.** In garden peas, one pair of alleles controls the height of the plant and a second pair of alleles controls flower color. The allele for tall (*D*) is dominant to the allele for dwarf (*d*), and the allele for purple (*P*) is dominant to the allele for white (*p*). A tall plant with purple flowers crossed with a dwarf plant with white flowers produces 1/4 tall purple, 1/4 tall white, 1/4 dwarf purple, and 1/4 dwarf white. What is the genotype of the parents?

D   **103.** In garden peas, one pair of alleles controls the height of the plant and a second pair of alleles controls flower color. The allele for tall (*D*) is dominant to the allele for dwarf (*d*), and the allele for purple (*P*) is dominant to the allele for white (*p*). A tall plant with white flowers crossed with a dwarf plant with purple flowers produces 1/4 tall purple, 1/4 tall white, 1/4 dwarf purple, and 1/4 dwarf white. What is the genotype of the parents?

D **104.** In garden peas, one pair of alleles controls the height of the plant and a second pair of alleles controls flower color. The allele for tall (D) is dominant to the allele for dwarf (d), and the allele for purple (P) is dominant to the allele for white (p). A tall plant with purple flowers crossed with a tall plant with white flowers produces 3/8 tall purple, 1/8 tall white, 3/8 dwarf purple, and 1/8 dwarf white. What is the genotype of the parents?

D **105.** In garden peas, one pair of alleles controls the height of the plant and a second pair of alleles controls flower color. The allele for tall (D) is dominant to the allele for dwarf (d), and the allele for purple (P) is dominant to the allele for white (p). A tall purple crossed with a tall purple produces 3/4 tall purple and 1/4 tall white. What is the genotype of the parents?

D **106.** In horses, black coat color is influenced by the dominant allele (B), and chestnut coat color is influenced by the recessive allele (b). Trotting gait is due to a dominant gene (T), pacing gait to the recessive allele (t). If a homozygous black trotter is crossed to a chestnut pacer,
   (a) what will be the appearance of the F_1 and F_2 generations?
   (b) which phenotype will be the most common?
   (c) which genotype will be the most common?
   (d) which of the potential offspring will be certain to breed true?

D **107.** In horses, black coat color is influenced by the dominant allele (B) and chestnut coat color by the recessive allele (b). Trotting gait is due to a dominant gene (T), pacing gait to the recessive allele (t). What color horse would you use to find out the genotype of a black trotter? Give the genotype and phenotype.

D **108.** Crosses between a yellow rat with a yellow rat always produce yellow. Crosses between a white rat with a white rat always produce white. The alleles affect the same aspect of coat color. The crosses of a white with a yellow produce a cream. What happens if you cross two creams?

D **109.** Assume red plants crossed with white plants give rise to pink plants. Explain how to eliminate red plants if you start with two pinks.

D **110.** If long or round are homozygous forms of an incompletely dominant gene and oval is the phenotype of the heterozygote, give the F_2 ratio of the cross between long and round (both genotype and phenotype).

D **111.** A breeder of cattle has a herd of white cows and a roan bull. Hair color in this breed is controlled by an incompletely dominant gene. The two homozygous forms are either red or white, and the heterozygous is roan.
   (a) What color of calves are expected and in what proportions?
   (b) Outline a procedure to develop an all-red herd.

D **112.** In radishes, two incompletely dominant genes control color and shape. Red and white radishes are homozygous, while the hybrid is purple. Long and round are homozygous and, if crossed, will produce an oval hybrid. Give the F_2 genotypic and phenotypic ratio produced by crossing pure-breed red long radishes with white round varieties.

D **113.** In a certain breed of chicken an incompletely dominant gene controls color. The homozygous black, when crossed with the homozygous splashed-white, produces an intermediate gray color pattern referred to as blue. A second gene controls the shape of the comb. The dominant allele (R) produces rose, while the recessive allele (r) produces single. Give the F_1 and F_2 genotypic and phenotypic ratios of a cross between a pure-breeding black single and a pure-breeding splashed-white rose.

D **114.** There are three alleles controlling the ABO blood types. I^A and I^B are codominant genes so that the combination $I^A I^B$ produces the AB blood type. The third allele I^O is recessive to the other two alleles. Indicate which of these parents could produce the given child:

	Parents	Child	Yes or No
(a)	A x AB	B	
(b)	A x O	A	
(c)	A x B	O	
(d)	A x AB	O	
(e)	A x AB	B	
(f)	B x B	O	
(g)	AB x AB	A	

D 115. In horses there are four alleles at the A locus. Arranged in dominance sequence they are:

A (wild) a^b (bay) a^c (brown) a^d (black)

If you bred several bay mares whose sires were brown to a brown stallion whose sire was black, what type of offspring would be produced and in what proportion?

D 116. In rabbits there are four alleles at the c locus. Arranged in dominance sequence they are:

C (agouti) c^{ch} (chinchilla) c^h (Himalayan) and c (albino)

(a) Is it possible to cross two agouti rabbits and produce both chinchilla and Himalayan offspring?

(b) Is it possible to cross two chinchillas and produce 1/2 chinchilla and 1/2 Himalayan?

D 117. Gray is homozygous while blue is a heterozygous form of a semi-lethal gene. Give the ratio of the offspring produced in the cross of two blues.

D 118. A cross of two Kerry horses always produces Kerry. A cross of a Kerry with a Dexter produces 1/2 and 1/2. Crosses of two Dexters produce two Dexters for every Kerry. Explain.

D 119. In the late 1920s, a mutation occurred in many silver fox farms around the world. The fox farms that sold expensive furs were proud of the quality of their furs, and each advertised that it had the best, most pure breed of all the fox farms. The new mutations produced a "platinum" coat pattern that was commercially desirable, so the farms crossed them to get more. The results of their breeding experiments were as follows: (1) silver x silver ——> all silver offspring; (2) silver x platinum ——> equal numbers of silver and platinum; (3) platinum x platinum ——> 2 platinum for each silver offspring. Explain.

D 120. There is a color pattern inherited in certain mice in which agouti (gray) is homozygous and yellow is heterozygous. A cross of two yellows produces two yellows for each agouti. A second gene, C/c, controls the expression of the color genes: C is the dominant allele that allows color to be expressed, and the recessive gene in the homozygous condition (cc) prevents any color from being expressed.

(a) Give the genotypes of a white parent crossed with a yellow parent that produces 1/2 white, 1/3 yellow, 1/6 agouti offspring.

(b) Give the results of the cross of two forms heterozygous for each gene.

(c) Can you develop a pure-breed population for any of the colors?

D 121. In poultry, the genes for rose comb (R) and pea comb (P) produce walnut whenever they occur together ($R\_ P\_$); single-combed individuals have the homozygous condition for both genes ($rr\ pp$).

(a) Give the F_1 and F_2 phenotypic results of a cross of a pure-breeding rose comb ($RR\ pp$) with a pure-breeding pea comb ($rr\ PP$).

(b) Give the phenotypic results of a cross of $Rr\ Pp$ x $rr\ Pp$.

(c) Give the phenotypic results of a cross of $RR\ Pp$ x $rr\ Pp$.

(d) Give the phenotypic results of a cross of $Rr\ pp$ x $rr\ Pp$.

(e) Give the phenotypic results of a cross of $Rr\ Pp$ x $rr\ pp$.

D 122. Congenital deafness in humans is due to the homozygous condition of either or both of the recessive genes d or e. Both dominant D and E are necessary for normal hearing. Gene D/d affects the middle ear, while gene E/e affects the inner ear. It does not matter how good the normal inner ear (as indicated by $E\_$) is; if there is something wrong in the middle ear, the individual is unable to hear. The same applies for the other gene. Give the phenotypic results of the following crosses:

(a) $Dd\ EE$ x $Dd\ EE$

(b) $Dd\ Ee$ x $Dd\ Ee$

(c) $dd\ EE$ x $DD\ ee$

(d) $Dd\ EE$ x $Dd\ ee$

(e) $Dd\ EE$ x $DD\ Ee$

D **123.** White fruit color in summer squash is influenced by a dominant allele W, while colored fruit must be ww. In the presence of ww, a dominant gene G results in yellow fruit, and if the individual had both recessive genes in the homozygous condition, it would be green. Give the F_2 phenotypic ratios resulting from a cross of a pure-breeding white of genotype $WW\ GG$ with a green.

D **124.** In cultivated stocks, the cross of a variety of white flower plants produced all red flowers in the F_1 generation, but the F_2 generation produced 87 red, 31 cream, and 39 white. Explain these results by giving the genotypes possible for each phenotype.

D **125.** In summer squash, spherical-shaped fruit has been shown to be dominant to elongated fruit. On one occasion two different spherical varieties were crossed and produced all disk-shaped fruits. When these hybrid disk-shaped fruits were crossed they produced 75 disk-shaped fruits, 48 spherical fruits, and 9 elongated fruits. Explain these results.

D **126.** In sweet peas, genes C and P are necessary for colored flowers. In the absence of either (__ pp or cc __), or both ($cc\ pp$), the flowers are white. What will be the color of the offspring of the crosses in what proportions for the following?
  (a) $Cc\ Pp$ x $cc\ pp$
  (b) $Cc\ Pp$ x $Cc\ Pp$
  (c) $Cc\ PP$ x $Cc\ pp$
  (d) $Cc\ pp$ x $cc\ Pp$

D **127.** In sweet peas, genes C and P are necessary for colored flowers. In the absence of either (__ pp or cc __), or both ($cc\ pp$), the flowers are white. Give the probable genotype of a plant with colored flowers and a plant with white flowers that produced 38 plants with colored flowers and 42 plants with white flowers.

D **128.** In a certain variety of plants a cross between a red-flowered plant and a white-flowered plant produced an all-red flower F_1. In the F_2 there were 140 red, 50 cream, and 65 white.
  (a) Offer an explanation for this F_2 ratio.
  (b) What ratio would be produced in a testcross of the F_1 hybrid?
  (c) What ratio would be produced if all the white F_2 plants were crossed among themselves?

D **129.** In a certain breed of chicken two genes control color. A dominant allele (I) inhibits the expression of any color gene (C). A second recessive gene (c) results in albinism when homozygous (cc). Give the F_2 phenotypic ratio of a colored chicken $ii\ CC$ with a white $II\ cc$.

D **130.** In mice the allele for colored fur (C) is dominant to the allele for albinism (c). The allele (W) for normal behavior is dominant to that for waltzing movement (w). Give the probable genotypes of the parents if they produced the offspring listed after the following crosses:
  (a) Colored normal x white waltzer produced 10 colored normal, 8 colored waltzers, 2 white waltzers, 11 white normal.
  (b) Colored normal x white normal produced 35 colored normal, 13 colored waltzers.
  (c) Colored normal x colored normal produced 37 colored normal, 14 colored waltzers, 9 white normal, and 5 white waltzers.

D **131.** Pure-breeding yellow guinea pigs crossed with pure-breeding white ones produce only cream-colored offspring. This pattern indicates incomplete dominance. Rough hair is found to be dominant to smooth hair. Give the F_1 and F_2 genotypic and phenotypic ratios of a cross of a smooth white guinea pig with a homozygous rough yellow guinea pig.

D **132.** There are nine coat colors known in foxes. If a red fox were crossed with a double-black fox, all the hybrids would be red above and black below in a pattern known as blended cross. If two blended crosses were mated, the F_2 ratio would be as follows: 1 red, 2 smokey red, 2 cross red, 4 blended cross, 1 standard silver, 2 substandard silver, 1 Alaskan silver, 2 sub-Alaskan silver, and 1 double black.

(a) Using the letters A/a and B/b to serve as the genes for these animals, develop a genotype for each variety listed.

(b) Two crosses will produce all blended-cross offspring. One is used above (red fox x double black); what is the other?

(c) List the genotype and phenotype of all the pure-breeding foxes.

(d) Give the genotypic and phenotypic ratio of a cross between two substandard silvers.

(e) Give the genotype and phenotype of the offspring produced in a cross of 1 sub-Alaskan silver and a cross red.

D **133.** In the garden pea Mendel found that tall (D) green pods (G) and inflated pods (C) were dominant to their alleles, dwarf (d) yellow pods (g) and constricted pods (c). Given the following genotypes, determine the chances of producing the offspring shown.

(a) $DD\ Gg\ Cc$ x $Dd\ Gg\ cc$ ——> $DD\ gg\ Cc$

(b) $DD\ Gg\ Cc$ x $Dd\ Gg\ Cc$ ——> tall green pod, constricted pod

(c) $Dd\ Gg\ Cc$ x $Dd\ GG\ cc$ ——> tall green pod, inflated pod

(d) $Dd\ Gg\ Cc$ x $Dd\ Gg\ Cc$ ——> $D\_\ G\_\ cc$

(e) $Dd\ Gg\ Cc$ x $Dd\ gg\ CC$ ——> $D\_\ G\_\ C\_$

(f) $Dd\ gg\ cc$ x $DD\ Gg\ cc$ ——> tall green pod, inflated pod

(g) $Dd\ Gg\ Cc$ x $Dd\ Gg\ Cc$ ——> $Dd\ Gg\ Cc$

(h) $Dd\ Gg\ Cc$ x $Dd\ Gg\ Cc$ ——> $dd\ gg\ cc$

D **134.** In tomatoes red (R) is dominant to yellow (r), tall (D) is dominant to dwarf (d), and smooth (H) is dominant to peach or hairy (h).

(a) How many different genotypes are there in relationship to these three characteristics?

(b) How many different phenotypes are there in relationship to these three characteristics?

(c) How many different homozygous pure-breeding forms can be produced?

D **135.** If you were following the inheritance patterns of two different sets of multiple alleles located on different chromosomes, how many different possible gametes could be produced if locus 1 had five possible alleles and locus 2 had six alleles?

Classification Questions

Answer questions 136–140 using the group of answers below.

a. 4
b. 6
c. 8
d. 12
e. 24

D **136.** In a dihybrid cross between a parent that is a double heterozygote ($Aa\ Bb$) and a parent that is homozygous dominant for one gene and heterozygous for the other ($AA\ Bb$), how many unique genotypes potentially will be present in their offspring?

D **137.** In a dihybrid cross between a parent that is a double heterozygote ($Aa\ Bb$) and a parent that is homozygous recessive for one gene and heterozygous for the other ($aa\ Bb$), how many unique phenotypes potentially will be present in their offspring?

D **138.** In a dihybrid cross between a parent that is a double heterozygote ($Aa\ Bb$) and a parent that is a double homozygous recessive ($aa\ bb$), how many unique phenotypes potentially will be present in their offspring?

D **139.** Plant species X is diploid ($2n = 24$) and has a quantitative trait, the expression of which is controlled by gene loci on each of its chromosomes. What is the maximum number of alleles for this trait that any one individual of species X could have?

D **140.** Animal species X is tetraploid ($4n = 12$). Following gene duplication and translocation, a given gene is found on each chromosome. How many alleles for this gene can be present in an individual of this species?

Answers: 136. b   137. a   138. a
           139. e   140. d

Selecting the Exception

D **141.** Four of the five answers listed below are dominant traits. Select the exception.
   a. green pod
   b. purple flower
   c. yellow seed coat
 * d. dwarf plant
   e. axial flower position

E **142.** Four of the five answers listed below describe the heterozygous condition. Select the exception.
 * a. homozygous
   b. carrier
   c. heterozygotes
   d. hybrid
   e. *Aa*

M **143.** Four of the five answers listed below describe the gene makeup. Select the exception.
   a. pure breeding
   b. homozygous
   c. heterozygous
   d. carrier
 * e. phenotype

M **144.** Four of the five answers listed below are accepted as valid explanations of genetic behavior. Select the exception.
 * a. blending
   b. dominance
   c. segregation
   d. independent assortment
   e. probability

E **145.** Four of the five answers listed below are pure breeding. Select the exception.
   a. *AA BB*
 * b. *Aa BB*
   c. *AA bb*
   d. *aa BB*
   e. *aa bb*

ANSWERS TO PROBLEMS IN CHAPTER 18

91. Red
92. Recessive
93. 1 *DD*, 2 *Dd*, 1 *dd*; 3 tall, 1 dwarf
94. Smooth hair, *ww*
95. (a) *Rr*, rose
     (b) *Rr* x *RR* ——> all rose, *Rr* x *rr* ——> 1/2 rose, 1/2 single
96. Black male (*Bb*) x brown female (*bb*)
     offspring: black (*Bb*) brown (*bb*)
97. (a) *Ss* x *Ss* ——> *S_* + *ss*
     (b) *Ss* x *ss* ——> *Ss*
     (c) *Ss* x *ss* ——> *Ss* + *ss*
     (d) *SS* x *S_* ——> *S_*
     (e) *ss* x *ss* ——> *ss*
98. 1/4 chance
99. Normal pigmented woman, *Cc*; albino father, *cc*; albino man, *cc*; normal parents, *Cc* + *Cc*; 3 children, 2 normal *Cc*, 1 albino *cc*
100. *DD Pp* x *dd pp*
101. *DD pp* x *dd PP*
102. *Dd Pp* x *dd pp*
103. *Dd pp* x *dd Pp*
104. *Dd Pp* x *Dd pp*
105. *Dd Pp* x *DD Pp*
106. (a) F$_1$: black trotters; F$_2$: 9 black trotters, 3 black pacers, 3 chestnut trotters, 1 chestnut pacer
     (b) Black pacer
     (c) *Bb Tt*
     (d) *bb tt*, chestnut pacers
107. *bb tt*, chestnut pacer
108. 1 yellow, 2 cream, 2 white
109. Cross until you get white, and use white in crosses until you cross two whites, then all subsequent plants will be white.
110 1 *LL*, 2 *Ll*, 1 *ll*; 1 long, 2 oval, 1 round

111. (a)   1/2 white, 1/2 roan

     (b)   Roan with white ——> roan;

         roan x roan ——> red;

         roan x red ——> red; red x red ——> red

112. 1 $LL\,RR$ long red, 2 $LL\,Rr$ long purple,

    2 $Ll\,RR$ oval red,

    4 $Ll\,RR$ oval purple, 1 $ll\,RR$ round red,

    1 $ll\,Rr$ round oval,

    1 $Ll\,rr$ long white, 2 $Ll\,rr$ oval white,

    1 $ll\,rr$ round white

113. F_1: $Bb\,Rr$ blue rose; F_2: 3 $BB\,R\_$ black rose,

    6 $Bb\,R\_$ blue rose, 3 $bb\,R\_$ splashed-white rose,

    1 $BB\,rr$ black single, 2 $Bb\,rr$ blue single,

    1 $bb\,rr$ splashed-white single

114. (a) yes; (b) yes; (c) yes; (d) no; (e) yes; (f) yes; (g) yes

115. $a^b\,a^c$ x $a^c\,a^d$; 1/2 bay, 1/2 brown

116. (a) no; (b) not likely but possible—would expect a 3:1

117. 1 gray, 2 blues (1 lethal)

118. Kerry is homozygous (DD), Exter is heterozygous (Dd), dd is lethal .

119. PP (silver), Pp (platinum), pp (lethal)

120. (a) $Yy\,Cc$ x $Yy\,cc$

     (b) $Yy\,Cc$ x $Yy\,Cc$ ——> 3/12 $YY\,C\_$ agouti, 6/12 $Yy\,C\_$ yellow, 3/12 $\_cc$ albino

     (c)   white and agouti

121. (a) F_1: $Rr\,Pp$ walnut; F_2: 9 $R\_P\_$ walnut, 3 R_pp rose, 3 $rrP\_$ pea, 1 $rrpp$ single

     (b)   3/8 walnut, 3/8 pea, 1/8 rose, 1/8 single

     (c)   3/4 walnut, 1/4 rose

     (d)   1/4 walnut, 1/4 rose, 1/4 pea, 1/4 single

     (e)   1/4 walnut, 1/4 rose, 1/4 pea, 1/4 single

122. (a)   3/4 normal, 1/4 deaf

     (b)   9/16 normal, 7/16 deaf

     (c)   all normal

     (d)   3/4 normal, 1/4 deaf

     (e)   all normal

123. 12/16 $W\_\_\_$ white, 3/16 $ww\,G\_$ yellow, 1/16 $ww\,gg$ green

124. R-red, r-cream, A-pigment, a-albino

    9 $R\_A\_$ red, 3 $rr\,A\_$ cream, 4 $\_\,aa$ white

125. 9 $D\_S\_$ disk, 3 D_ss spherical, 3 $dd\,S\_$ spherical, 1 $dd\,ss$ elongated; presence of both dominant genes produces disk, while the presence of either one of the genes as homozygous recessive produces spherical, and both recessive produces elongated fruit.

126. (a)   1/4 color, 3/4 white

     (b)   9/16 color, 7/16 white

     (c)   3/4 color, 1/4 white

     (d)   1/4 color, 3/4 white

127. $Cc\,Pp$ x $CC\,pp$

128. (a) R-red, r-cream, A-color, a-albino; 9 $R\_A\_$ red, 3 $rr\,A\_$ cream, 4 $\_\_\,aa$ white

     (b)   1/4 red, 1/4 cream, 1/2 white

     (c)   all offspring would be white because all would be $\_\_\,aa$

129. 9 $I\_C\_$ + 3 $I\_\,cc$ + 1 $ii\,cc$ = 13 white + 3 $ii\,C\_$ = 3 color

130. (a) $Cc\,Ww$ x $cc\,ww$

     (b) $CC\,Ww$ x $cc\,Ww$

     (c) $Cc\,Ww$ x $Cc\,Ww$

131. F_1: rough cream $Yy\,Rr$; F_2: 3 $YY\,R\_$ yellow rough, 6 $Yy\,R\_$ cream rough, 3 $yy\,R\_$ white rough, 1 $YY\,rr$ yellow smooth, 2 $Yy\,rr$ cream smooth, 1 $yy\,rr$ white smooth

132. (a)   1 $AA\,BB$ red

         2 $AA\,Bb$ smokey red

         2 $Aa\,BB$ cross red

         4 $Aa\,Bb$ blended cross

         1 $aa\,BB$ standard silver

         2 $aa\,Bb$ substandard silver

         1 $AA\,bb$ Alaskan silver

         2 $Aa\,bb$ sub-Alaskan silver

         1 $aa\,bb$ double black

     (b) $aa\,BB$ standard silver x $AA\,bb$ Alaskan silver

     (c) $AA\,BB$ red, $AA\,bb$ Alaskan silver, $aa\,BB$ standard silver, $aa\,bb$ double black

     (d)   1 standard silver $aa\,BB$

         2 substandard silver $aa\,Bb$, 1 double black $aa\,bb$

     (e)   1 smokey red $AA\,Bb$, 2 blended cross $Aa\,Bb$, 1 substandard silver $aa\,Bb$

133. (a)   (1/2) (1/4) (1/2) = 1/16

     (b)   (1) (3/4) (1/4) = 3/16

     (c)   (3/4) (1) (1/2) = 3/8

     (d)   (3/4) (3/4) (1/4) = 9/64

     (e)   (3/4) (1/2) (1) =3/8

     (f)   (1) (1/2) 0 = 0

     (g)   (2/4) (2/4) (2/4) = 8/64

     (h)   (1/4) (1/4) (1/4) = 1/64

134. (a)   27   (b)   8   (c)   8

135. 30 possibilities

CHAPTER 19

CHROMOSOME VARIATIONS AND MEDICAL GENETICS

Multiple-Choice Questions

THE CHROMOSOMAL BASIS OF INHERITANCE

E **1.** Genes are
   a. located on chromosomes.
   b. inherited in the same way as chromosomes.
   c. arranged in linear sequence on chromosomes.
   d. assorted independently during meiosis.
   * e. all of these

M **2.** Which of the following is NOT true concerning homologous chromosomes?
   a. There two of each kind.
   b. Each parent contributes one of each homologous pair.
   c. Most homologous carry the same genes for the same traits.
   * d. The number of homologous chromosomes is doubled in each generation.
   e. Homologous chromosomes pair up during early meiosis.

E **3.** Chromosomes other than those involved in sex determination are known as
   a. nucleosomes.
   b. heterosomes.
   c. alleles.
   * d. autosomes.
   e. liposomes.

M **4.** DNA coding regions that affect the same trait are called
   a. homologues.
   * b. alleles.
   c. autosomes.
   d. loci.
   e. gametes.

E **5.** The location of a gene on a chromosome is its
   a. centromere.
   * b. locus.
   c. autosome.
   d. allele.

E **6.** A karyotype
   a. compares one set of chromosomes to another.
   * b. is a visual display of chromosomes arranged according to size.
   c. is a photograph of cells undergoing mitosis during anaphase.
   d. of a normal human cell shows 48 chromosomes.
   e. cannot be used to identify individual chromosomes beyond the fact that two chromosomes are homologues.

E **7.** In karyotyping, individual chromosomes may be distinguished from others by
   a. a comparison of chromosome lengths.
   b. bands produced on chromosomes by differential staining.
   c. the position of centromeres.
   * d. all of these

M **8.** Karyotyping is usually done using what kind of cells?
   a. muscle
   * b. blood
   c. cartilage
   d. sex
   e. epidermal

M **9.** Which chemical is used to keep chromosomes from separating during metaphase?
   a. Giemsa stain
   b. acetone
   * c. colchicine
   d. alcohol
   e. formaldehyde

M **10.** Karyotyping involves taking pictures of chromosomes during
   a. prophase.
   b. telophase.
   * c. metaphase.
   d. interphase.
   e. anaphase.

M   11. Karyotype analysis
      a. is a means of detecting and reducing mutagenic agents.
      b. is a surgical technique that separates chromosomes that have failed to segregate properly during meiosis II.
      * c. is used in prenatal diagnosis to detect chromosomal mutations and metabolic disorders in embryos.
      d. substitutes defective alleles with normal ones.
      e. all of these

D   12. With respect to chromosomes, the difference between normal human males and females is defined by which of the following?
      a. In females, one X is deleted.
      b. Females possess one X and one Y.
      * c. In males an X is replaced by a Y.
      d. Females have three X's.
      e. Males have two X's and a Y.

E   13. Sex chromosomes
      a. determine gender.
      b. vary from one sex to another.
      c. carry some genes that have nothing to do with sex.
      d. were unknown to Mendel.
      * e. all of these

D   14. Concerning the sex chromosomes, which of the following is correct?
      a. The Y chromosome carries a greater number of nonsexual traits.
      b. X and Y are different in size but carry nearly equal numbers of genes.
      * c. The X chromosome carries more genes for nonsexual traits.
      d. The X chromosome carries only gender-related genes.
      e. The X chromosome carries the TDF gene.

E   15. Which of the following is an accurate characterization of a mutation?
      a. an exchange of chromosomes between two chromosomes
      b. the linkage of two unrelated chromosomes
      * c. a change in the nucleotides of DNA.
      d. the reassortment of chromosomes at meiosis
      e. the shuffling of genes during gamete preparation

SEX DETERMINATION AND X INACTIVATION

D   16. Which of the following statements is false?
      a. The SRY gene is absent in all females.
      b. The SRY gene apparently is the gene that controls the development of male sexuality.
      * c. The development of maleness is by default because males lack 2 X chromosomes.
      d. Maleness develops in the embryo before femaleness.
      e. There is no difference in external genitalia of males or females until four weeks after conception when the genes determining sex begin to be expressed.

E   17. In human females one of the sex chromosomes is switched off during early development in a phenomenon called
      a. karyotyping.
      * b. X inactivation.
      c. X linkage.
      d. crossing over.
      e. Sry activation.

E   18. A condensed female X chromosome is called a
      * a. Barr body.
      b. Morgan sphere.
      c. Sry gene.
      d. karyotype.
      e. linkage map.

LINKED GENES AND CROSSING OVER

M   19. All of the genes located on a given chromosome comprise a
      a. karyotype.
      b. bridging cross.
      c. wild-type allele.
      * d. linkage group.

E   20. If two genes are on the same chromosome,
      a. crossing over occurs frequently.
      b. they assort independently.
      * c. they are in the same linkage group.
      d. they are segregated during meiosis.
      e. an inversion will usually occur.

M   21. If two genes are almost always found in the same gamete,
     *   a. they are located close together on the same chromosome.
          b. they are located on nonhomologous chromosomes.
          c. they are located far apart on the same chromosome.
          d. they are found on the sex chromosome.
          e. all except "they are located far apart on the same chromosome."

M   22. Genes that are located on the same chromosome
          a. tend to be inherited together.
          b. will appear together in the gamete.
          c. are said to be linked.
          d. may be separated during crossing over.
     *   e. all of these

D   23. If alleles *L*, *M*, and *N* are on the maternal chromosome and *l*, *m*, and *n* are on the paternal chromosome, the only way that a gamete from a heterozygote will produce a gamete with alleles *l*, *m*, and *N* is through
          a. nondisjunction.
          b. the laws of segregation.
          c. the law of independent assortment.
     *   d. crossing over.
          e. chromosome aberration.

D   24. If the paternal chromosome has alleles *L*, *M*, and *n* and the maternal chromosomes have *l*, *m*, and *N*, then the chromosome that cannot be produced by crossing over is
          a. *LMN*
     *   b. *LMn*
          c. *LmN*
          d. *Lmn*
          e. *lmn*

M   25. Genetic recombination as a result of crossing over occurs more readily in genes that
          a. are on the sex chromosomes.
          b. are on the autosomes.
          c. are located close together on the same chromosome.
     *   d. are located far apart on the same chromosome.
          e. are located on different chromosomes.

D   26. Gene mapping
          a. applies only to genes located on the same chromosomes.
          b. represents actual physical distance between genes.
     *   c. is based upon the frequency of crossing over.
          d. can be accomplished only by using the sex chromosomes.

D   27. Which of the following statements is false?
          a. Crossing over tends to reduce the frequency that two linked genes are inherited together.
          b. Independent assortment of homologous chromosomes during meiosis increases variation.
          c. Crossing over leads to variation.
          d. Abnormal number or structure of chromosomes may influence the course of evolution.
     *   e. The closer together genes are found on a chromosome the greater is the chance that crossing over will occur between them.

D   28. In genetic analyses, researchers know that linkage of genes will introduce exceptions to the principle of
          a. dominance.
          b. segregation.
          c. recessiveness.
     *   d. independent assortment.
          e. chromosomal inheritance.

D   29. Which of the following is (are) correct?
          a. The ability to map the position of genes on a chromosome depends on the frequency of crossing over.
          b. The map distances between genes on a chromosome map mark only relative distances between them, depending on crossover frequency.
          c. The chance of a crossover only partially depends upon the distance between genes.
          d. The map distance in a chromosome map refers to actual physical distance between genes.
     *   e. both "The ability to map the position of genes on a chromosome depends on the frequency of crossing over;" and "The map distances between genes on a chromosome map mark only relative distances between them, depending on crossover frequency."

M   **30.** Linkage mapping is a technique that determines

     a. the relationship between chromatids.

   *   b. the positions of genes on chromosomes relative to one another.

     c. the positions of chiasmata.

     d. the sex of offspring.

     e. the probability of a deletion.

HUMAN GENETIC ANALYSIS

E   **31.** In a pedigree chart a male showing the specific trait being studied is indicated by a

   *   a. darkened square.

     b. clear square.

     c. darkened diamond.

     d. clear triangle.

     e. darkened circle.

E   **32.** In a pedigree chart a female who does not demonstrate the trait being studied is represented by a

     a. darkened square.

     b. clear diamond.

   *   c. clear circle.

     d. darkened triangle.

     e. darkened oval.

E   **33.** When the results of some genetic condition seriously affects the individual, it is called

     a. an abnormality.

   *   b. a disorder.

     c. a deviation.

     d. a carrier.

     e. a pedigree.

M   **34.** Amniocentesis involves sampling

     a. the fetus directly.

   *   b. the fetal cells floating in the amniotic fluid.

     c. sperm.

     d. blood cells.

     e. placental cells.

M   **35.** Amniocentesis is

     a. a surgical means of repairing deformities.

     b. a form of chemotherapy that modifies or inhibits gene expression or the function of gene products.

   *   c. used in prenatal diagnosis to detect chromosomal mutations and metabolic disorders in embryos.

     d. a form of gene replacement therapy.

     e. all of these

M   **36.** The most recent technique for analyzing the genetics of the unborn child involves the sampling of

     a. the fetus directly.

     b. cells in the amniotic fluid.

     c. material from the allantois.

   *   d. the chorionic villi.

     e. yolk sac material.

M   **37.** Gene replacement therapy

     a. has not yet been used successfully with mammals.

     b. is a surgical technique that separates chromosomes that have failed to segregate properly during meiosis II.

     c. has been used successfully to treat victims of Huntington's disorder by removing the dominant damaging autosomal allele and replacing it with a harmless one.

   *   d. substitutes defective alleles with normal ones.

     e. all of these

PATTERNS OF AUTOSOMAL INHERITANCE

E   **38.** Symptoms of phenylketonuria (PKU) may be minimized or suppressed by a diet low in

     a. serine.

     b. glycine.

   *   c. phenylalanine.

     d. proline.

     e. glutamic acid.

D   **39.** An autosomal recessive disorder

     a. requires that only one parent be a carrier.

     b. displays its symptoms only in heterozygotes.

     c. is more frequent in males than females.

   *   d. will appear only in children of parents who both carry the gene.

     e. is dominant in females.

M   **40.** The probability of producing a normal child by two parents who are carriers for an autosomal recessive disorder is

     a. 50 percent.

     b. 0 percent.

     c. 100 percent.

     d. 25 percent.

   *   e. 75 percent.

M 41. The probability of producing a child who suffers from cystic fibrosis by two parents who are carriers for the autosomal recessive disorder is
    a. 50 percent.
    b. 0 percent.
    c. 100 percent.
  * d. 25 percent.
    e. 75 percent.

D 42. If a study of several pedigrees demonstrated that two parents are normal but their children express a trait, then the trait is controlled by a
    a. codominant gene.
    b. simple dominant gene.
  * c. recessive gene.
    d. sex-linked gene.
    e. No conclusion can be drawn.

D 43. A woman is diagnosed to have the genetic disease known as Huntington's disorder. It is a rare defect caused by an autosomal dominant allele. The chance for any one of her children to inherit the disease is
    a. dependent on the sex of the child.
    b. 1 out of 3.
  * c. 1 out of 2
    d. 3 out of 4

D 44. In an autosomal dominant disorder such as Huntington's, two carrier parents have the probability of passing the gene on to _____ percent of their children.
    a. 50
    b. 0
    c. 100
    d. 25
  * e. 75

D 45. Tay-Sachs disease
    a. is controlled by a simple recessive gene.
    b. is a sex-linked disease more common in males.
  * c. occurs only in those individuals that receive two copies of a defective recessive gene.
    d. is the result of the failure of chromosomes to separate so that an individual receives 3 instead of 2 chromosomes.

E 46. In which of the following does the onset of symptoms usually occur after child bearing age?
    a. Tay-Sachs
    b. hemophilia
  * c. Huntington's
    d. muscular dystrophy
    e. achondroplasia

M 47. Familial hypercholesterolemia is a condition explainable by
  * a. autosomal dominance.
    b. X linkage.
    c. autosomal recessiveness.
    d. translocation.
    e. chromosomal duplication.

PATTERNS OF X-LINKED INHERITANCE

D 48. A color-blind man and a woman with normal vision whose father was color blind have a son. Color blindness, in this case, is caused by an X-linked recessive gene. If only the male offspring are considered, the probability that their son is color blind is
    a. 25 percent.
  * b. 50 percent.
    c. 75 percent.
    d. 100 percent.
    e. none of these

M 49. Red-green color blindness is an X-linked recessive trait in humans. A color-blind woman and a man with normal vision have a son. What is the probability that the son is color blind?
  * a. 100 percent
    b. 75 percent
    c. 50 percent
    d. 25 percent
    e. 0 percent

M 50. Red-green color blindness is an X-linked recessive trait in humans. What is the probability that a color-blind woman and a man with normal vision will have a color-blind daughter?
    a. 100 percent
    b. 75 percent
    c. 50 percent
    d. 25 percent
  * e. 0 percent

M   **51.** If a daughter expresses an X-linked recessive gene, she inherited the trait from
- a. her mother.
- b. her father.
- * c. both parents.
- d. neither parent.
- e. her grandmother.

M   **52.** A human X-linked recessive gene may be
- a. found on the Y chromosome.
- b. passed to daughters from their fathers.
- c. passed to sons from their mothers.
- d. expressed more commonly among females.
- * e. both passed to daughters from their fathers and passed to sons from their mothers.

M   **53.** An X-linked carrier is a
- a. homozygous dominant female.
- * b. heterozygous female.
- c. homozygous recessive female.
- d. homozygous male.
- e. heterozygous male.

M   **54.** A human X-linked gene is
- a. found only in males.
- b. more frequently expressed in females.
- c. found on the Y chromosome.
- d. transmitted from father to son.
- * e. found on the X chromosome.

D   **55.** Color blindness is an X-linked trait in humans. If a color-blind woman marries a man with normal vision, the children will be
- a. all color-blind daughters, but normal sons.
- * b. all color-blind sons, but carrier daughters.
- c. all normal sons, but carrier daughters.
- d. all color-blind children.
- e. all normal children.

M   **56.** A woman heterozygous for color blindness (an X-linked recessive allele) marries a man with normal color vision. What is the probability that their first child will be color blind?
- * a. 25 percent
- b. 50 percent
- c. 75 percent
- d. 100 percent
- e. none of these

E   **57.** Queen Victoria
- * a. was a carrier of hemophilia.
- b. had a hemophilic parent.
- c. had hemophilia.
- d. married a man with hemophilia.
- e. both had a hemophilic parent and married a man with hemophilia.

M   **58.** Hemophilia
- a. is rare in the human population.
- b. is more common among men.
- c. was common in English royalty.
- d. is an X-linked recessive trait.
- * e. all of these

M   **59.** Which of the following would be considered a carrier of a sex-linked recessive defect?
- a. a man with the defect
- b. a woman with the defect
- c. a father of a son with the defect
- * d. the normal daughter whose father had the defect
- e. a son of two unaffected parents

D   **60.** Males tend to be affected in greater numbers by X-linked recessive genetic disorders than are females because
- a. females have two dominant genes for the disorder.
- * b. males have only one recessive gene for the disorder.
- c. males have a double dose of the gene.
- d. Y chromosomes are not as strong as X chromosomes.
- e. females have two dominant genes for the disorder and males have only one recessive gene for the disorder.

SEX-INFLUENCED INHERITANCE

E   **61.** Pattern baldness is a trait that is referred to as sex
- a. linked.
- b. related.
- c. retarded.
- * d. influenced.
- e. matched.

CHANGES IN CHROMOSOME STRUCTURE

M   **62.** Which is NOT a chromosomal aberration?
   a. deletion
   b. extra chromosomes
   c. translocation (exchange of parts between non homologs)
   * d. crossing over
   e. inversion

M   **63.** A chromosome's gene sequence that was ABCDEFG before modification and ABCDLMNOP afterward is an example of
   a. inversion.
   b. deletion.
   c. duplication.
   * d. translocation.
   e. crossing over.

M   **64.** A chromosome's gene sequence that was ABCDEFG before modification and ABCDCDEFG afterward is an example of
   a. inversion.
   b. deletion.
   * c. duplication.
   d. translocation.
   e. crossing over.

E   **65.** A chromosome that has been broken and rejoined in a reversal sequence has undergone
   * a. inversion.
   b. deletion.
   c. duplication.
   d. translocation.
   e. crossing over.

E   **66.** A chromosome's gene sequence that was ABCDEFG before damage and ABCFG after is an example of
   a. inversion.
   * b. deletion.
   c. duplication.
   d. translocation.
   e. crossing over.

E   **67.** A chromosome's gene sequence that was ABCDEFG before damage and ABFEDCG after is an example of
   * a. inversion.
   b. deletion.
   c. duplication.
   d. translocation.
   e. crossing over.

M   **68.** Certain human cancer cells may demonstrate which of the following?
   a. deletion
   b. inversion
   * c. translocation
   d. duplication
   e. none of these

E   **69.** A chromosome that has been broken and rejoined in a reversal sequence has undergone
   * a. inversion.
   b. deletion.
   c. duplication.
   d. translocation.
   e. aneuploidy.

M   **70.** Which of the following is a transfer of genes between nonhomologous chromosomes?
   a. crossing over
   b. aneuploidy
   c. trisomy
   * d. translocation
   e. duplication

CHANGES IN CHROMOSOME NUMBER

M   **71.** The condition occurring when an organism has a $2n + 1$ chromosome composition is known as
   a. monosomy.
   * b. trisomy.
   c. diploid.
   d. haploid.
   e. both trisomy and haploid.

E   **72.** If a gamete is missing one chromosome,
   a. the chromosome number is expressed as $2n - 1$.
   b. then one chromosome is without its homologue.
   c. the condition is called monosomy.
   d. the chromosome number is expressed as $2n - 1$ and the condition is called monosomy.
   * e. the chromosome number is expressed as $2n - 1$, the condition is called monosomy, and one chromosome is without its homologue.

D   **73.** If nondisjunction occurs during meiosis,
- a. the resulting sex cells will be heterogametes.
- * b. one-half of the resulting cells will exhibit trisomy and the other half monosomy.
- c. diploid cells will be produced.
- d. all gametes would lack a chromosome and these gametes would be infertile.

E   **74.** The failure of chromosomes to separate during mitosis or meiosis is called
- a. genetic displacement.
- b. trisomy.
- c. crossing over.
- * d. nondisjunction.
- e. disjunction.

D   **75.** Which of the following is different from the other four?
- * a. nondisjunction
- b. duplication
- c. inversion
- d. deletion
- e. translocation

D   **76.** Aneuploidy would describe all of the following except
- a. Turner syndrome.
- b. Klinefelter syndrome.
- * c. translocation.
- d. XYY.
- e. Down syndrome.

E   **77.** Which of the following designates a normal human female?
- a. XXY
- b. XY
- * c. XX
- d. XYY
- e. XO

E   **78.** Which of the following designates a normal human male?
- a. YY
- b. XX
- * c. XY
- d. XO
- e. XYY

D   **79.** Suppose a hemophilic male (X-linked recessive allele) and a female carrier for the hemophilic trait have a nonhemophilic daughter with Turner syndrome. Nondisjunction could have occurred in
- a. both parents.
- b. neither parent.
- * c. the father only.
- d. the mother only.

E   **80.** Down syndrome involves trisomy
- a. 3.
- b. 5.
- c. 15.
- d. 19.
- * e. 21.

E   **81.** Syndrome means
- a. a chromosome disorder.
- b. a simple genetic disease.
- * c. a set of symptoms that occur together.
- d. an incurable disease.
- e. a rare inborn defect.

M   **82.** In Down syndrome
- * a. as the age of the mother increases, the chance of the defect occurring in the unborn children increases.
- b. the father usually has less influence on the defect.
- c. most embryos abort before complete term.
- d. a person with the defect cannot have a normal child.
- e. none of these

M   **83.** A genetic abnormality that may result in sterile males with mental retardation or breast enlargement is
- * a. XXY.
- b. XYY.
- c. Turner syndrome.
- d. Down syndrome.
- e. XXX

E   **84.** The sex chromosome composition of a person with Klinefelter syndrome is most accurately written as
- a. XXX.
- b. XO.
- * c. XXY.
- d. XYY.
- e. Y

M   85. Males who tend to be taller than average and show mild mental retardation may be designated
       a. XXY.
    * b. XYY.
       c. Turner syndrome.
       d. Down syndrome.
       e. XXX

D   86. Nondisjunction involving the X chromosomes may occur during oogenesis and produce two kinds of eggs. If normal sperm fertilize these two types, which of the following pairs of genotypes are possible?
       a. XX and XY
    * b. XXY and XO
       c. XYY and XO
       d. XYY and YO

E   87. The sex chromosome composition of a person with Turner syndrome is most accurately written as
       a. XXX.
    * b. XO.
       c. XXY.
       d. XYY.
       e. X

D   88. The chromosome composition of metafemales is
       a. XO.
       b. XX.
    * c. XXX.
       d. XXY.
       e. XYY.

D   89. Which of the following conditions is characterized by a karyotype with 45 chromosomes?
    * a. Turner syndrome
       b. Down syndrome
       c. testicular feminization syndrome
       d. Klinefelter syndrome
       e. cri-du-chat

Matching Questions

M   90. Matching I

1 ____   colchicine
2 ____   deletion
3 ____   duplication
4 ____   inversion
5 ____   monosomy
6 ____   translocation
7 ____   trisomy

A. a chromosome segment is permanently transferred to a nonhomologous chromosome

B. $(2n - 1)$; a gamete deprived of a chromosome

C. a repeat of a particular DNA sequence in the same chromosome or in nonhomologous ones

D. $(2n + 1)$; three chromosomes of the same kind are present in a set of chromosomes

E. a piece of the chromosome is inadvertently left out during the repair process

F. inhibits microtubule assembly; prevents chromosome movement

G. a chromosome segment that has been cut out and rejoined at the same place, but backward

Answers: 1. F   2. E   3. C
              4. G   5. B   6. A
              7. D

E   91. Matching II. Match the cause with the disorder.

1 ____   Down syndrome
2 ____   phenylketonuria
3 ____   hemophilia
4 ____   Turner syndrome

A. autosomal recessive inheritance; phenylalanine accumulates

B. nondisjunction of the twenty-first chromosomal pair

C. X-linked recessive inheritance

D. nondisjunction of the sex chromosomes

Answers: 1. B   2. A   3. C
              4. D

Problems

D **92.** In cats the allele B produces black, while b produces yellow. Neither gene is dominant, and in the heterozygous state the phenotype is a combination of yellow and black spots called tortoiseshell. The alleles B and b are X-linked. If a tortoiseshell cat has three tortoiseshell kittens and two black kittens, give the genotype and phenotype of the tomcat that produced them, give the sex of the kittens.

D **93.** An X-linked recessive gene (c) produces red/green color blindness. A normal woman whose father was color blind marries a color-blind man.
  a. What are the possible genotypes for the mother of the color-blind man?
  b. What are the possible genotypes for the father of the color-blind man?
  c. What are the chances that the first son will be color blind?
  d. What are the chances that the first daughter will be color blind?

D **94.** In cats an X-linked pair of alleles, B and b, controls color of fur. The alleles are incompletely dominant: B produces black, b produces yellow, and Bb produces tortoise-shell.
  a. A yellow cat had a litter of two tortoiseshell kittens and one yellow. What is the sex of the yellow kitten?
  b. A tortoiseshell cat brings home a litter of black, yellow, and tortoiseshell kittens. The color of which sex would tell you the color of the tomcat that produced them?
  c. A yellow male is crossed with a tortoiseshell female. If the female has all male kittens in her litter of four, what color(s) would they be?
  d. A tortoiseshell cat brings home her litter of black, yellow, and tortoiseshell kittens. By what method could you possibly decide whether the male parent was the black tomcat next door?

D **95.** If a father and a son are both color blind and the mother is normal, is it likely that the son inherited color blindness from his father?

D **96.** If a human recessive X-linked characteristic occurred with a 10 percent frequency, what would its frequency be in males and females?

D **97.** In humans an X-linked disorder called coloboma iridia (a fissure in the iris) is a recessive trait. A normal couple has an afflicted daughter. The husband sues the wife for divorce on the grounds of infidelity. Would you find in his favor?

D **98.** In *Drosophila* a narrow reduced eye is called a bar-eye. It is due to a dominant X-linked allele (B), while the full wild-type is due to the recessive gene ($B+$). Give the F_1 and F_2 genotypic and phenotypic expectations of a cross of a homozygous wild-type female with a bar-eyed male.

D **99.** If the gene for yellow body color (y) is an X-linked recessive and its dominant counterpart ($y+$) produces wild body colors, give the phenotypes expected and their frequencies for the following four crosses:
  a. yellow female x wild male
  b. wild carrier female x wild male
  c. wild carrier female x yellow male
  d. homozygous wild female x yellow male

D **100.** Two *Drosophila* are crossed several times, with a total number of offspring of 106 females and 48 males. There is too great a deviation from the expected 1:1 ratio for chance alone to account for the difference. What other factor could account for this difference?

D **101.** White eyes in *Drosophila* is a mutation that turned out to be an X-linked recessive. Would you expect that the first time the white eye was discovered it was in a male or female?

D **102.** Hemophilia is an X-linked recessive gene. A normal woman whose father had hemophilia marries a normal man. What are the chances of hemophilia in their children?

D **103.** Color blindness is an X-linked recessive gene. Two normal-visioned parents produce a color-blind child.
  a. Is this child male or female?
  b. What are the genotypes of the parents?
  c. What are the chances that their next child will be a color-blind daughter?

D **104.** If an X-linked recessive gene is expressed in 4 percent of the men, what proportion of women would express the recessive trait?

D **105.** Red/green color blindness is an X-linked recessive trait. Two normal-visioned parents have a color-blind son. Indicate the genotype and phenotype of each parent and the son.

D **106.** There is an autosomal gene that controls baldness, and its expression is sex influenced, so that the gene for baldness (B) is dominant in males but recessive in females. In females the allele B^1 for nonbaldness is dominant over the gene for baldness. If a heterozygous nonbald woman marries a nonbald man, what will be the appearance of their children? Work out the possibilities for each sex.

D **107.** Short index fingers (shorter than ring finger) are dominant in males and recessive in females, while long index fingers (as long or longer than ring fingers) are dominant in females and recessive in males. Give the F_2 genotype and phenotype resulting from the cross of a male with long fingers with a female with short fingers.

Classification Questions

Answer questions 108-112 in reference to the five items listed below.

    a. 12
    b. 23
    c. 24
    d. 46
    e. 47

D **108.** How many chromosomes does each somatic cell have in a human male who has two X chromosomes?

D **109.** Following a gene duplication event involving only five loci, how many chromosomes will a human female have?

D **110.** How many chromosomes are present in the somatic cells of a child born with Down syndrome (trisomy 21)?

D **111.** How many chromosomes are present in each cell of the germ cell line for a tetraploid species where its normal complement of chromosomes is 48?

D **112.** The normal sperm cell of a particular species carries 11 chromosomes. Following nondisjunction in the formation of secondary spermatocytes and their subsequent fertilization of normal ova, some of the zygotes will have 21 chromosomes, others will have 22, and the remainder will have how many chromosomes?

Answers: 108. e   109. d   110. e
                111. c   112. b

Answer questions 113-117 in reference to the five processes listed below. Note that to answer these questions you need to know that the sequence of amino acids directly reflects the sequence of genes that coded for their placement.

    a. an inversion
    b. a deletion
    c. a gene duplication
    d. a translocation
    e. an addition

D **113.** Homologous sets of genes ABCDEF and aBCdEF are located on nonhomologous chromosomes. Crossing over between them is suppressed because their locations are the result of ____?

D **114.** Homologous sets of genes ABCDEF and AEDCBF are located on homologous chromosomes. Crossing over between them is suppressed because of ____?

D **115.** A small region of a protein from three species is sequenced and found to be as follows:

    species X is alanine, glycine, glycine, threonine, alanine
    species Y is alanine, glycine, threonine, alanine
    species Z is alanine, valine, glycine, threonine, alanine

The difference in the amino acid sequence of species Y is most likely due to ____?

D   **116.** A small region of a protein from three species is sequenced and found to be as follows:

> species X is alanine, valine, threonine, alanine
>
> species Y is alanine, glycine, threonine, alanine
>
> species Z is alanine, valine, glycine, threonine, alanine

The differences in the amino acid sequence of species Z is most likely due to ____?

D   **117.** The nucleotide sequences of homologous regions of DNA of two species is AATGCCCCGTTA and AATGCCCCGCTTA. If this is not the result of a nucleotide base-pair addition, then it is most likely the result of ____?

Answers: 113. d   114. a   115 b
           116. e   117. b

Answer questions 118-122 in reference to the five disorders listed below:

     a. cystic fibrosis
     b. Turner syndrome
     c. AIDS
     d. hemophilia
     e. Down syndrome

D   **118.** Gene therapy has been tried recently for this disorder.

D   **119.** This disorder is an autosomal recessive disorder.

M   **120.** This disorder is an X-linked recessive trait.

E   **121.** This disorder is also known as trisomy 21.

D   **122.** This disorder is due to a sex chromosome abnormality probably caused by nondisjunction of sex chromosomes at meiosis.

Answers: 118. a   119. a   120. d
           121. e   122. b

Selecting the Exception

D   **123.** Four of the five answers listed below provide evidence that genes are located on chromosomes. Select the exception.
     a. the chromosome number is cut in half by meiosis
     b. original chromosome number restored by fertilization
     c. some genes tend to be inherited together
*   d. environmental factors may influence gene expression
     e. there are two sets of chromosomes, one maternal, one paternal in diploid forms

M   **124.** Four of the five answers listed below are related conditions in which abnormal numbers of chromosomes are present. Select the exception.
     a. monosomy
     b. Down syndrome
     c. nondisjunction
*   d. complete chromosome set
     e. trisomy

M   **125.** Four of the five answers listed below are conditions caused by chromosomal nondisjunction. Select the exception.
     a. Down syndrome
*   b. Huntington's disorder
     c. Turner syndrome
     d. Klinefelter syndrome
     e. trisomy 21

D   **126.** Four of the five answers listed below are caused by recessive genes. Select the exception.
*   a. Huntington's disorder
     b. phenylketonuria
     c. color blindness
     d. hemophilia
     e. albinism

ANSWERS TO PROBLEMS IN CHAPTER 19

92. *B Y* black, tortoiseshell female, black males

93. a. *Cc* or *cc*   b. *C Y* or *c Y*
    c. 1/2   d. 1/2

94. a. male   b. female   c. 1/2 yellow 1/2 black
    d. black female kitten

95. No, males inherit all sex-linked traits from the mother.

96. Males 10 percent, females 1 percent

97. Yes, the daughter would have to inherit the recessive trait from both parents.

98. *B+B+* x *B Y* ——> F_1: *B+B* wild female
    B+Y wild male F_2: 1/4 *B Y* bar male
    1/4 *B+Y* wild male 1/4 *B+B+* wild female
    1/4 *B+B* wild female

99. a. yellow male wild female
    b. 1/4 wild male 1/4 yellow male
       1/4 wild female 1/4 wild carrier female
    c. 1/4 wild male 1/4 yellow male
       1/4 yellow female 1/4 wild carrier female
    d. 1/2 wild carrier female 1/2 wild male

100. a sex-linked recessive lethal gene expressed in the males, who received it from their mothers

101. Male

102. All females normal but 1/2 of them would be carriers; 1/2 of the males would have hemophilia, the other 1/2 normal.

103. a. male   b. *Cc* x *C Y*
     c. no chance to produce a color-blind daughter

104. 4/100 x 4/100 = 16/10,000, or 4 out of 2,500

105. father *C Y*, mother *Cc*, son *c Y*

106. BB^1 x B^1B^1 ——> BB^1 + B^1B^1; all daughters nonbald, 1/2 sons bald, 1/2 sons nonbald

107. F_2: 3/4 males with short fingers, 1/4 males with long fingers F_2: 3/4 females with long fingers, 1/4 females with short fingers

CHAPTER 20

DNA STRUCTURE AND FUNCTION

Multiple-Choice Questions

DNA STRUCTURE

E   **1.** The building blocks of nucleic acids are
     a. amino acids.
*   b. nucleotides.
     c. pentose sugars.
     d. phosphate groups.
     e. nitrogenous bases.

M   **2.** In the pairing of two nucleotides within the double helix
     a. hydrogen bonds are used.
     b. adenine and thymine bind together.
     c. cytosine binds with guanine.
*   d. all of these

E   **3.** A nucleotide may contain a
     a. base.
     b. 5-carbon sugar.
     c. phosphate group.
     d. adenine.
*   e. all of these

D   **4.** Which of the following terms is NOT related to the other four?
*   a. amino acids
     b. nucleotides
     c. pentose sugars
     d. phosphate groups
     e. nitrogenous bases

M   **5.** James Watson and Francis Crick
     a. established the double-stranded nature of DNA.
     b. established the principle of base pairing.
     c. explained how DNA's structure permitted it to be replicated.
     d. proposed the concept of the double-helix.
*   e. all of these

E   **6.** In the bonding of nitrogenous bases
     a. adenine is paired with cytosine.
     b. adenine is paired with guanine.
     c. cytosine is paired with thymine.
*   d. guanine is paired with cytosine.
     e. two of the above

E   **7.** The DNA molecule could be compared to a
     a. hairpin.
*   b. ladder.
     c. key.
     d. globular mass.
     e. flat plate.

E   **8.** In DNA, complementary base pairing occurs between
     a. cytosine and uracil.
     b. adenine and guanine.
     c. adenine and uracil.
*   d. adenine and thymine.
     e. all of these

M   **9.** Adenine and guanine are
*   a. nitrogenous bases.
     b. base pairs bonded together in DNA
     c. bases found only in RNA
     d. waste products of protein synthesis.
     e. amino acids.

E   **10.** In DNA molecules
*   a. the nucleotides are arranged in a linear, unbranched pattern.
     b. the nitrogenous bases are found on the outside of the molecule.
     c. the pentose-phosphate pattern runs the same way on each DNA strand.
     d. all of these
     e. none of these

M   **11.** Which statement is true?
*   a. The hydrogen bonding of cytosine to guanine is an example of complementary base pairing.
     b. Adenine always pairs up with guanine in DNA, and cytosine always teams up with thymine.
     c. Each of the four nucleotides in a DNA molecule has the same nitrogen-containing base.
     d. When adenine base pairs with thymine, they are linked by three hydrogen bonds.
     e. In the DNA of all species, the amount of cytosine never equals the amount of guanine

E   12. Each DNA strand has a backbone that
         consists of alternating
         a. amino acid units.
         b. nitrogen-containing bases.
         c. hydrogen bonds.
    *   d. sugar and phosphate molecules.
         e. amines and purines.

E   13. The DNA molecule usually is made up of
         how many strands?
             a. 1
    *   b. 2
             c. 3
             d. 6
             e. 12

M   14. In the comparison between a spiral staircase
         and a DNA molecule, the steps would
         correspond to
             a. sugars.
             b. hydrogen bonds.
    *.   c. base pairs.
             d. nucleotides.
             e. phosphates.

M   15. Biochemically, a gene is BEST defined as a
         a. unit of heredity.
         b. part of a chromosome.
    *   c. region of DNA that codes for protein
            assembly.
         d. portion of the nuclear chromatin.
         e. part of the material located in the
            nucleus.

M   16. A linear stretch of DNA that specifies the
         sequence of amino acids in a polypeptide is
         called a(n)
             a. codon.
             b. intron.
             c. messenger.
    *   d. gene.
             e. enzyme.

M   17. Each DNA strand serve as which of the
         following during DNA synthesis?
             a. replicate
             b. substitute
    *   c. template
             d. source of nucleotides
             e. all of these

DNA REPLICATION AND REPAIR

M   18. The appropriate adjective to describe DNA
         replication is
             a. nondisruptive.
    *   b. semiconservative.
             c. progressive.
             d. natural.
             e. lytic.

D   19. The ultimate explanation for resemblances
         of traits from one generation to another is
             a. gamete formation.
    *   b. semiconservative DNA replication.
             c. sexual reproduction.
             d. protein synthesis.
             e. bloodlines.

E   20. Replication of DNA
             a. produces RNA molecules.
             b. produces only new DNA.
    *   c. produces two molecules each of which
                is half-new and half-old DNA joined
                lengthwise to each other.
             d. generates excessive DNA, which
                eventually causes the nucleus to divide.
             e. is too complex to characterize.

M   21. DNA polymerase
             a. is an enzyme.
             b. adds new nucleotides to a strand.
             c. proofreads DNA strands to see that they
                are correct.
             d. derives energy from ATP for synthesis
                of DNA strands.
    *   e. all of these

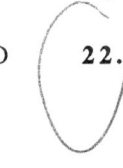

D   22. After three replications of a single DNA
         molecule, what percent of the resulting
         double helices contain one strand of the
         "original" DNA?
             a. 0 percent
    *   b. 25 percent
             c. 50 percent
             d. 75 percent
             e. 100 percent

DNA INTO RNA—PROTEIN SYNTHESIS BEGINS

E   23. The RNA molecule is made up of how
         many strands?
    *   a. 1
             b. 2
             c. 3
             d. 6
             e. 12

E   **24.** In terms of their nitrogenous base component, how many different kinds of RNA molecules are there?

       a. 3

*   b. 4

       c. 5

       d. 6

       e. 12

E   **25.** The form of RNA that carries the code from the DNA to the site where the protein is assembled is called

*   a. messenger RNA.

       b. nuclear RNA.

       c. ribosomal RNA.

       d. transfer RNA.

       e. structural RNA.

M   **26.** Which of the following carries amino acids to ribosomes, where amino acids are linked into the primary structure of a polypeptide?

       a. mRNA

*   b. tRNA

       c. hnRNA

       d. rRNA

       e. all of these

M   **27.** Transfer RNA differs from other types of RNA because it

       a. transfers genetic instructions from cell nucleus to cytoplasm.

       b. specifies the amino acid sequence of a particular protein.

*   c. carries an amino acid at one end.

       d. contains codons.

       e. none of these

M   **28.** The central dogma of molecular biology

       a. explains the structural complexity of genes.

*   b. describes the flow of information.

       c. is based upon the role of proteins in controlling life.

       d. does not explain how genes function.

       e. explains evolution in terms of molecular biology.

D   **29.** The changing of a business letter from shorthand to typewritten copy is analogous to

       a. translation of mRNA.

*   b. transcription of DNA.

       c. protein synthesis.

       d. deciphering the genetic code.

       e. replication of DNA.

E   **30.** All the different kinds of RNA are transcribed in the

       a. mitochondria.

       b. cytoplasm.

       c. ribosomes.

*   d. nucleus.

       e. endoplasmic reticulum.

E   **31.** The nitrogenous base found in DNA but not in RNA is

       a. adenine.

       b. cytosine.

       c. guanine.

       d. uracil.

*   e. thymine.

E   **32.** Which substance is found in RNA but not in DNA?

       a. thymine

       b. deoxyribose

*   c. ribose

       d. guanine

       e. cytosine

E   **33.** The nitrogenous base found in RNA but not in DNA is

       a. adenine.

       b. cytosine.

       c. guanine.

*   d. uracil.

       e. thymine.

D   **34.** DNA and RNA are alike in

       a. the pentose sugar.

       b. all the nitrogenous bases used to assemble the genetic code.

       c. the number of strands.

       d. their function in genetics.

*   e. none of these

E   **35.** Uracil will pair with

       a. ribose.

*   b. adenine.

       c. cytosine.

       d. thymine.

       e. guanine.

M   **36.** The synthesis of an RNA molecule from a DNA template strand is

       a. replication.

       b. translation.

*   c. transcription.

       d. DNA synthesis.

       e. metabolism.

E   37. The relationship between strands of RNA and DNA is
    a. antagonistic.
    b. opposite.
  * c. complementary.
    d. an exact duplicate.
    e. unrelated.

M   38. Transcription
    a. occurs on the surface of the ribosome.
    b. is the final process in the assembly of a protein.
  * c. occurs during the synthesis of any type of RNA from a DNA template.
    d. is catalyzed by DNA polymerase.
    e. all of these

M   39. Which of the following dominates in the process of transcription?
  * a. RNA polymerase
    b. DNA polymerase
    c. phenylketonuria
    d. transfer RNA
    e. all of these

M   40. Transcription
    a. involves both strands of DNA as templates.
    b. uses the enzyme DNA polymerase.
    c. results in a double-stranded end product.
  * d. produces three different types of RNA molecules.
    e. all of these

M   41. Transcription starts at a region of DNA called a(n)
    a. sequencer.
  * b. promoter.
    c. activator.
    d. terminator.
    e. transcriber.

M   42. In transcription
    a. several RNA molecules are made from the same DNA molecule.
    b. promoters are needed so that RNA polymerase can bind to DNA.
    c. DNA produces messenger RNA.
    d. a specific enzyme called RNA polymerase is required.
  * e. all of these

M   43. The portion of the DNA molecule that is translated is composed of
    a. introns.
    b. anticodons.
  * c. exons.
    d. transcriptions.
    e. both exons and transcriptons.

M   44. The portion of the DNA molecule that is not translated and is a noncoding portion of DNA is composed of
  * a. introns.
    b. anticodons.
    c. exons.
    d. transcriptions.
    e. both exons and transcriptons.

M   45. In transcription
    a. several amino acids are assembled by the messenger RNA molecules at one time.
  * b. a special sequence called a promoter is necessary for transcription to begin.
    c. certain polypeptide sequences are governed by one ribosome, while other sequences are produced by other ribosomes.
    d. the transfer RNA molecules arrange the messenger RNA codons into the appropriate sequence.
    e. none of these

FROM mRNA TO PROTEINS

M   46. When a gene transcription occurs, which of the following is produced?
    a. more DNA
    b. protein or polypeptide sequences
  * c. messenger RNA
    d. enzymes
    e. genetic defects

E   47. The genetic code is made up of units each of which consists of how many nucleotides?
    a. 2
  * b. 3
    c. 5
    d. 6
    e. 12

E **48.** There are how many different kinds of
amino acids in proteins?
　　a. 3
　　b. 6
　　c. 12
＊　d. 20
　　e. 28

M **49.** There are how many different kinds of RNA
codons?
　　a. 3
　　b. 12
　　c. 28
＊　d. 64
　　e. 120

M **50.** If the codon consisted of only two
nucleotides, there would be how many
different kinds of codons?
　　a. 4
　　b. 8
＊　c. 16
　　d. 32
　　e. 64

M **51.** The concept that a set of three nucleotides
specifies a particular amino acid provides
the basis for
　　a. the one gene, one enzyme hypothesis.
　　b. the one gene, one polypeptide
　　　hypothesis.
＊　c. the genetic code.
　　d. biochemical reactions among nucleic
　　　acids.
　　e. all of these

M **52.** Of all the different codons that exist, three
of them
　　a. are involved in mutations.
　　b. do not specify a particular amino acid.
　　c. cannot be copied.
　　d. provide punctuation or instructions
　　　such as "stop."
＊　e. do not specify a particular amino acid
　　　but do provide punctuation or
　　　instructions such as "stop."

E **53.** Each "word" in the mRNA language
consists of how many letters?
＊　a. three
　　b. four
　　c. five
　　d. more than five
　　e. none of these

D **54.** If each nucleotide coded for a single amino
acid, how many different types of amino
acids could be combined to form proteins?
＊　a. four
　　b. sixteen
　　c. twenty
　　d. sixty-four
　　e. none of these

M **55.** Which of the following carries amino acids
to ribosomes, where amino acids are linked
into the primary structure of a polypeptide?
　　a. mRNA
＊　b. tRNA
　　c. hnRNA
　　d. rRNA
　　e. all of these

E **56.** Ribosomes function as
　　a. a single unit.
＊　b. two-part units.
　　c. three-part units.
　　d. four-part units.
　　e. a multidivisional unit.

HOW IS mRNA TRANSLATED?

D **57.** All mRNA transcripts begin with
　　a. methionine.
　　b. a ribosome.
＊　c. AUG.
　　d. the P site.
　　e. an anticodon.

M **58.** A polysome is
　　a. one of the units of a ribosome.
　　b. the nuclear organelle that synthesizes
　　　RNA.
　　c. an organelle that functions similarly to
　　　a ribosome during meiosis.
　　d. the two units of a ribosome considered
　　　together.
＊　e. an mRNA molecule with several
　　　ribosomes attached.

M **59.** If the DNA triplets were ATG-CGT, the
mRNA codons would be
　　a. AUGCGU.
　　b. ATGCGT.
＊　c. UACGCA.
　　d. UAGCGU.
　　e. none of these

M   **60.** If the DNA triplets were ATG-CGT, the tRNA anticodons would be
* a. AUGCGU.
  b. ATGCGT.
  c. UACGCA.
  d. UAGCGU.
  e. none of these

E   **61.** In what stage of translation does the peptide bond join two amino acids?
  a. initiation
  b. transcription
* c. elongation
  d. reparation
  e. termination

EFFECTS OF MUTATIONS

E   **62.** The difference between normal hemoglobin and sickle-cell hemoglobin is in the
  a. heme portion of the molecules.
  b. number of chains of amino acids.
* c. substitution of a specific amino acid for another specific amino acid.
  d. addition of one amino acid to the normal hemoglobin molecule.
  e. loss of only one amino acid from the normal hemoglobin molecule.

M   **63.** A gene mutation
  a. is a change in the nucleotide sequence of DNA.
  b. may be caused by environmental agents.
  c. may arise spontaneously.
  d. can occur in all organisms.
* e. all of these

E   **64.** Mutations can be
  a. random.
  b. beneficial.
  c. harmful.
  d. heritable.
* e. all of these

M   **65.** Frameshift mutations may involve
  a. substitution of nucleotides.
  b. substitution of codons.
  c. substitution of amino acids.
* d. addition or deletion of one to several base pairs.
  e. all of these

M   **66.** Sickle-cell anemia has been traced to what type of mutation?
  a. frameshift
  b. transposable element
  c. mutagenic
* d. base-pair substitution
  e. viral

M   **67.** In a mutation,
  a. the new codon may specify a different amino acid, but may not change the function of the new protein produced.
  b. the new codon may specify the same amino acid as the old codon.
  c. the new codon and resulting amino acid may destroy the function of the protein specified.
* d. All of these may be true.

E   **68.** A DNA region that can "jump" from one location to another is called a
  a. frameshift mutation.
* b. transposable element.
  c. mutagen.
  d. polysome.
  e. base-pair substitution.

M   **69.** Fragile X syndrome and Huntington disorder are examples of what type of mutation?
  a. frameshift
  b. base-substitution
* c. expansion
  d. transposable element
  e. point

REGULATING GENE ACTIVITY

D   **70.** Which of the following statements is false?
  a. All the cells of any one single individual human have the same genes.
* b. Only red blood cells have hemoglobin genes.
  c. Different genes are activated depending on the cell they are in.
  d. Genetically, liver cells and brain cells differ mostly in function.
  e. Cancer is the second highest cause of death by disease in the United States.

M   **71.** Genes located in different regions of the body during embryonic development may be
  a. turned on and off.
  b. never turned on.
  c. turned on and left on.
  d. activated for only a short time in one cell and a long time in another cell.
* e. all of these

D   72. The fact that prolactin stimulates milk glands but thyroxine does not is due to
   a. the inability of the thyroxine to reach the milk glands.
   b. the lack of ducts leading from the source of thyroxine production to the milk glands.
   c. chance.
 * d. the lack of proper receptors on the milk gland cells.
   e. pregnancy.

M   73. The target for prolactin is the
   a. pituitary gland.
   b. pancreas.
   c. parathyroid.
 * d. mammary gland.
   e. salivary gland.

Matching Questions

D   74. Matching. Choose the one most appropriate answer for each.
   1 ____ histone
   2 ____ promoter
   3 ____ intron
   4 ____ codon
   5 ____ transposable element
   6 ____ tRNA

   A. base sequences that are not translated into an amino acid sequence
   B. DNA region that can "jump" to another location
   C. proteins that help keep DNA molecules organized
   D. carries amino acid to protein assembly sites
   E. triplet of nucleotide bases
   F. base sequence that signals the start of a gene

Answers: 1. C   2. F   3. A
             4. E   5. B   6.D

D   75. Matching. Choose the best matching element.
   1 ____ anticodon
   2 ____ codon
   3 ____ messenger RNA
   4 ____ promoters
   5 ____ transcription
   6 ____ translation

   A. RNA-directed synthesis of polypeptide chains
   B. sites at which RNA polymerases can bind and initiate transcription
   C. binds to small subunit platform of a ribosome
   D. guided and catalyzed by RNA polymerases
   E. a tRNA triplet opposite an amino acid
   F. a set of three nucleotides

Answers: 1. E   2. F   3. C
             4. B   5. D   6. A

Classification Questions

Answer questions 76-80 in reference to the five nucleotides listed below:
   a. guanine
   b. cytosine
   c. adenine
   d. thymine
   e. uracil

E   76. Early data indicated that within a species the amount of adenine was always equal to the amount of this nucleotide.

E   77. This nucleotide is not incorporated into the structure of the DNA helix.

D   78. This nucleotide base pairs with cytosine.

M   79. This base in DNA would pair up with uracil in transcription.

M   80. Two hydrogen bonds connect adenine to _____ in the DNA molecule.

Answers: 76. d   77. e   78. a
             79. c   80. d

Classification Questions

Answer questions 81-85 in reference to the five RNA codons listed below:

        a. AUG
        b. UAA
        c. UUU
        d. UUA
        e. AAA

D   **81.** This codon terminates a coding region.

D   **82.** The anticodon AAA would pair with this codon.

M   **83.** A single mutation involving the second letter of codon AUA would convert it to this.

M   **84.** A DNA codon of ATT would be complementary to this RNA codon.

M   **85.** This codon codes for an amino acid and indicates the beginning of a coding region.

Answers: 81. b   82. c   83. e
            84. b   85. a

Selecting the Exception

M   **86.** Four of the five answers listed below are bases used to construct nucleic acids. Select the exception.
        a. cytosine
        b. adenine
        c. thymine
        d. guanine
   *   e. phenylalanine

D   **87.** Four of the five answers listed below are correctly paired. Select the exception.
   *   a. A - C
        b. C - G
        c. A - T
        d. T - A
        e. A - U

E   **88.** Three of the four answers listed below are different forms of a class of nucleic acids. Select the exception.
   *   a. template
        b. ribosomal
        c. messenger
        d. transfer

M   **89.** Three of the four answers listed below are involved in gene action. Select the exception.
        a. replication
        b. transcription
        c. translation
   *   d. polymerization

D   **90.** Four of the five answers listed below describe changes at the chromosomal level. Select the exception.
   *   a. base substitution
        b. duplication
        c. translocation
        d. deletion
        e. inversion

D   **91.** Four of the five answers listed below are chromosomal abnormalities. Select the exception.
        a. translocation
   *   b. elongation
        c. duplication
        d. deletion
        e. inversion

D   **92.** Four of the five answers listed below are components of a nucleotide. Select the exception.
        a. pentose sugar
   *   b. amino acid
        c. cytosine
        d. phosphate group
        e. adenine

D   **93.** Four of the five answers listed below are related by a common number. Select the exception.
        a. number of nucleotides in a codon
        b. number of building blocks (parts) in a nucleotide
        c. number of stop codons
   *   d. number of types of DNA
        e. number of types of RNA

D   **94** Three of the four answers listed below are steps in translation. Select the exception.
        a. initiation
   *   b. replication
        c. chain elongation
        d. termination

CHAPTER 21

RECOMBINANT DNA AND GENETIC ENGINEERING

Multiple-Choice Questions

A TOOLKIT FOR MAKING RECOMBINANT DNA

M 1. Recombinant DNA
   a. has occurred in sexually reproducing forms.
   b. can be produced with new biological techniques.
   c. occurs with viral infections of various forms of life.
   d. has produced changes that resulted in evolution.
   * e. all of these

E 2. New genetic combinations result from
   a. crossing over.
   b. sexual reproduction.
   c. mutations.
   d. exchange of genes between different species.
   * e. all of these

M 3. Recombinant DNA technology
   a. uses bacteria to make copies of the desired product.
   b. splices DNAs together.
   c. is possible only between closely related species.
   * d. uses bacteria to make copies of the desired product and splices DNAs together.
   e. uses bacteria to make copies of the desired product and splices DNAs together, but is possible only between closely related species.

E 4. Small circular molecules of DNA in bacteria are called
   * a. plasmids.
   b. desmids.
   c. pili.
   d. F particles.
   e. transferrins.

E 5. A tangelo is a combination
   a. orange and lemon.
   b. orange and tangerine.
   c. navel orange and tangerine.
   d. tangerine and cantaloupe.
   * e. tangerine and grapefruit.

D 6. Plasmids
   a. are self-reproducing circular molecules of DNA.
   b. are sites for inserting genes for amplification.
   c. may be transferred between different species of bacteria.
   d. may confer the ability to donate genetic material when bacteria conjugate.
   * e. all of these

E 7. Enzymes used to cut DNA molecules in recombinant DNA research are
   a. ligases.
   * b. restriction enzymes.
   c. transcriptases.
   d. DNA polymerases.
   e. replicases.

M 8. The fragments of chromosomes split by restriction enzymes
   a. have fused ends.
   b. have specific sequences of nucleotides.
   c. have sticky ends.
   d. form a circle.
   * e. have specific sequences of nucleotides and sticky ends.

M 9. The "natural" use of restriction enzymes by bacteria is to
   a. integrate viral DNA.
   * b. destroy viral DNA.
   c. repair "sticky ends."
   d. copy the bacterial genes.
   e. clone DNA.

M 10. Restriction enzymes
 a. often produce staggered cuts in DNA that are useful in splicing genes.
 b. are like most enzymes in being very specific in their action.
 c. are natural defense mechanisms evolved in bacteria to guard against or counteract bacteriophages.
 d. are used along with ligase and plasmids to produce a DNA library.
 * e. all of these

M 11. Restriction enzymes
 * a. work at recognition sites.
 b. function only at "sticky ends."
 c. produce uniform lengths of DNA.
 d. function only in genetic laboratories.
 e. none of these

M 12. Which of the following enzymes joins the paired sticky ends of DNA fragments?
 a. reverse transcriptase
 b. restriction enzymes
 * c. DNA ligase
 d. DNA polymerase
 e. transferase

D 13. Because it has no introns, researchers prefer to use _____ when working with human genes.
 * a. cDNA
 b. cloned DNA
 c. hybridized DNA
 d. RFLPs
 e. viral DNA

M 14. RNA can manufacture DNA via the action of
 a. DNA polymerase.
 b. RNA polymerase.
 * c. reverse transcriptase.
 d. ligase.
 e. restriction endonuclease.

PCR—A FASTER WAY TO AMPLIFY DNA

D 15. Which of the following methods of DNA amplification does NOT require cloning?
 a. reverse transcription
 * b. polymerase chain reaction
 c. cloned DNA
 d. reverse transcription and polymerase chain reaction
 e. polymerase chain reaction and cloned DNA

D 16. For polymerase chain reaction to occur,
 a. isolated DNA molecules must be primed.
 b. all DNA fragments must be identical.
 c. the DNA must be separated into single strands.
 d. a sticky end must be available for the ligase enzyme to function.
 * e. isolated DNA molecules must be primed and the DNA must be separated into single strands.

D 17. Multiple copies of DNA can be produced by
 a. cloning a DNA library.
 b. genetic amplification.
 c. the use of reverse transcriptase.
 d. the action of DNA polymerase.
 * e. all of these

SCIENCE COMES TO LIFE: DNA FINGERPRINTS

D 18. The use of RFLPs for "genetic fingerprinting" is based upon
 a. the type of gel used in electrophoresis.
 b. identical alleles at loci.
 * c. differences of locations where enzymes make their cuts.
 d. differences between blood and semen DNA.
 e. bonding of DNA to RNA.

D 19. Which of the following statements about restriction fragment length polymorphism is false?
 a. RFLPs can be used as a genetic fingerprint.
 b. RFLPs are based upon variations in alleles at the same locus.
 c. RFLPs reflect the fact that molecular differences in alleles alter the site where restriction enzymes function.
 * d. RFLPs can be used to distinguish between identical twins.
 e. RFLPs have greatly increased the number of sites involved in mapping the human genome.

HOW IS DNA SEQUENCED?

M   **20.** The detection of nucleotides as they pass through an automated DNA sequencing machine is by
    a. radioactivity.
*   b. laser flourescence.
    c. antibiotic resistance.
    d. electron microscopy.
    e. gel electrophoresis.

E   **21.** To determine the nucleotide sequence in DNA, scientists now use
    a. electron microscopy.
    b. antibiotic resistance.
*   c. automated DNA sequencing.
    d. radioactive tracers.
    e. viruses.

FROM HAYSTACKS TO NEEDLES— ISOLATING SPECIFIC GENES

M   **22.** A collection of DNA fragments produced by restriction enzymes and incorporated into plasmids is called
    a. copied DNA.
    b. transcribed DNA.
    c. DNA amplification.
*   d. a DNA library.
    e. plasmid DNA.

D   **23.** Probes for cloned genes use
*   a. complementary nucleotide sequences labeled with radioactive isotopes.
    b. certain media with specific antibodies.
    c. specific enzymes.
    d. certain bacteria sensitive to the genes.
    e. all of these

D   **24.** The method used to determine which host cells pick up a desired plasmid is the use of
    a. fluorescent dyes.
    b. restriction enzymes.
*   c. antibiotics.
    d. marker genes.
    e. a known series of nonsense nucleotides (introns).

ENGINEERING BACTERIA AND PLANTS

E   **25.** Genetically-engineered organisms that carry some foreign genes are said to be
    a. mutated.
    b. restricted.
*   c. transgenic.
    d. cloned.
    e. replicated.

M   **26.** Vaccines produced by genetic engineering contain
    a. plasmids in solution.
*   b. antigen produced by bacteria,
    c. weakened or killed microbes.
    d. antibodies.
    e. live viruses.

E   **27.** Genetically engineering bacteria may be used to break down pollutants in the environment in a process known as
*   a. bioremediation.
    b. replica plating.
    c. hybridization.
    d. transcription.
    e. polymorphism.

M   **28.** Which statement is true?
    a. There is no danger involved in recombinant DNA research in humans.
    b. There is no danger involved in recombinant DNA research in bacteria.
    c. There is no danger in releasing recombinant organisms into the environment.
*   d. Stringent safety rules make the use of recombinant DNA research possible.
    e. It is safe to conduct recombinant DNA research in plants.

GENE TRANSFERS IN ANIMALS

D   **29.** The human genome project seeks to
*   a. identify the nucleotide sequence of all human genes.
    b. develop a complete DNA library for a human gene.
    c. develop genetic markers for all genetic diseases.
    d. catalog all the varieties of human alleles.
    e. identify all humans that possess genetic defects.

M 30. Which of the following would represent a way to affect a cell without directly affecting its genes?
   a. Insert plasmids.
 * b. Inhibit transcription of mRNA.
   c. Use bioremediation.
   d. Activate reverse transcription.
   e. Speed up the action of DNA polymerase.

E 31. What is the name given to bits of RNA that act like enzymes to cut up specific mRNA sequences before they can be translated?
   a. restriction enzymes
   b. cloning vectors
   c. RNA ligase
   d. exons
 * e. ribozymes

METHODS AND PROSPECTS FOR GENE THERAPY

M 32. Gene therapy
   a. has not yet been used successfully with mammals.
   b. is a surgical technique that separates chromosomes that have failed to segregate properly during meiosis II.
   c. has been used successfully to treat victims of Huntington's disorder by removing the dominant damaging autosomal allele and replacing it with a harmless one.
 * d. offers the possibility of replacing defective alleles with normal ones.
   e. all of these

E 33. Recombinant DNA research uses plasmids and which of the following as cloning vectors?
   a. *E. coli*
 * b. viruses
   c. plants
   d. fungi
   e. any type of host cell

Classification Questions

Answer questions 34-38 in reference to the five items listed below:
   a. restriction enzymes
   b. recombinants
   c. plasmids
   d. clones
   e. restriction sites

M 34. These are bacterial populations containing thousands or millions of identical copies of one to several genes.

D 35. When one uses the techniques of genetic engineering to move a novel or foreign piece of DNA into the DNA of an organism, these new DNA regions are produced.

M 36. The pieces of DNA that are moved by a genetic engineer from one organism to another are first incorporated these.

E 37. The sole function of these is to cut apart foreign DNA molecules.

D 38. This is the designation of the specific nucleotide sequences where restriction endonucleases cleave DNA.

Answers: 34. d   35. b   36. c
            37. a   38. e

Answer questions 39-43 in reference to the four items listed below:
   a. cDNA
   b. a restriction enzyme
   c. reverse transcriptase
   d. a DNA library

M 39. This is from a viral source and catalyzes reactions to construct DNA strands from mRNA.

E 40. Any DNA copied from mRNA transcripts is known as this.

E 41. This a nuclease whose only function is to cut apart foreign DNA entering a cell.

M 42. Collections of DNA fragments produced by restriction enzymes and incorporated into cloning vectors are known as this.

D 43. This is a type of bacterial colony probe constructed of radioactively labeled DNA subunits.

Answers: 39. c   40. a   41. b
            42. d   43. a

Selecting the Exception

M   44. Four of the five answers listed below are aspects of the process known as gene splicing. Select the exception.
    a. cloning vector
    b. restriction enzymes
    c. sticky ends
    d. exposed base pairs
*   e. crossing over

M   45. Four of the five enzymes below are used in genetic engineering. Select the exception.
    a. ligase
    b. reverse transcriptase
    c. restriction
*   d. replicase
    e. DNA polymerase

M   46. Four of the five statements below are true of cloned DNA. Select the exception.
    a. The plasmid used is the cloning vector.
    b. Identical copies are produced.
*   c. Cloned DNA is produced by reverse transcriptase.
    d. Multiple copies are produced.
    e. Cloned DNA is manufactured in bacteria cells.

CHAPTER 22

CANCER: A CASE STUDY OF GENES AND DISEASE

Multiple-Choice Questions

WHAT IS CANCER?

M 1. Genes located in different regions of the body during embryonic development may be
   a. turned on and off.
   b. never turned on.
   c. turned on and left on.
   d. activated for only a short time in one cell and a long time in another cell.
   * e. all of these

E 2. Which of the following terms is most synonymous with "tumor?"
   a. cancer
   b. dysplasia
   * c. neoplasm
   d. oncogene
   e. metastasis

M 3. Which of the following would NOT be a characteristic of a benign tumor?
   a. enclosed by a capsule of connective tissue
   b. well differentiated
   c. slow growth
   * d. metastatic
   e. usually not life threatening

E 4. Under the microscope, cancerous tumors
   a. display well-differentiated cells.
   * b. have ragged edges.
   c. appear highly organized.
   d. are of normal cell size and appearance.
   e. all of these

E 5. Which term refers to the processes by which cells with identical genotypes become structurally and functionally distinct from one another?
   a. metamorphosis
   b. metastasis
   c. cleavage
   * d. differentiation
   e. induction

M 6. Cancer cells
   a. have altered plasma membranes.
   b. are unable to attach to other cells.
   c. divide to produce high densities of cells.
   d. have a different metabolism, using glycolysis even when oxygen is available.
   * e. all of these

E 7. Which of the following would NOT be characteristic of a cancer cell?
   * a. abnormally small nucleus
   b. abnormal cytoskeleton
   c. less cytoplasm than usual
   d. altered plasma membrane
   e. lack of strong cell-to-cell junctions

M 8. Cancerous growths appear to grow faster than the tissues around them because
   a. their mitosis rates are higher.
   * b. they do not stop dividing even when crowding occurs.
   c. the cancer cells inhibit the reproduction of the surrounding normal cells.
   d. their mitosis rates are higher and they do not stop dividing even when crowding occurs.
   e. their mitosis rates are higher; they do not stop dividing even when crowding occurs; and the cancer cells inhibit the reproduction of the surrounding normal cells.

M 9. The spread of a cancer from one site to others in the body is known as
   a. benign tumor.
   * b. metastasis.
   c. malignant tumor.
   d. remission.
   e. benign tumor and malignant tumor.

E 10. Cancer cells are able to stimulate the growth of blood vessels by secreting
   a. interleukin-2.
   b. p53.
   * c. angiogenin.
   d. aflatoxin.
   e. monoclonal antibodies.

THE GENETIC TRIGGERS FOR CANCER

E 11. Proto-oncogenes
   a. code for abnormal growth factors.
   b. are the precursors to malignancy.
 * c. operate in healthy cells.
   d. cannot be altered once they are inherited.
   e. code for abnormal growth factors and are the precursors to malignancy.

E 12. By analogy, a tumor suppressor gene would correspond to what part of an automobile?
   a. accelerator
   b. engine
   c. windshield
   d. fuel tank
 * e. brake

E 13. By analogy, an oncogene would correspond to what part of an automobile?
 * a. accelerator
   b. engine
   c. windshield
   d. fuel tank
   e. brake

D 14. A mutation in a tumor suppressor gene would most likely
   a. have no effect on cells.
   b. decrease the possibility of cancer.
   c. lead to the production of cancer suppressing proteins.
 * d. increase the possibility of cancer.
   e. change proto-oncogenes to oncogenes.

M 15. The p53 gene
   a. is a tumor suppressor gene.
   b. turns on proto-oncogenes.
   c. signals cells to stop division at the appropriate time.
   d. can mutate by a base substitution.
 * e. all of these

M 16. Which of the following is NOT a method by which an oncogene can be turned on?
   a. mutation
   b. translocation
 * c. deletion of the proto-oncogene
   d. viral invasion

E 17. The conversion of a normal cell into a cancerous one is called
 * a. carcinogenesis.
   b. metastasis.
   c. tumorization.
   d. biopsy.
   e. interferon.

M 18. Viruses can cause cancer in a host cell by
   a. disrupting the plasma membrane.
 * b. altering the DNA.
   c. slowing its metabolism.
   d. destroying the nucleus.
   e. removing gene sequences.

E 19. The specific name given to a cancer-producing chemical is
   a. pathogen.
 * b. carcinogen.
   c. teratogen.
   d. mutagen.
   e. oncogene.

E 20. For most people, the greatest radiation risk leading to cancer is
   a. medical and dental X rays.
   b. cosmic rays.
   c. radon gas.
 * d. sun exposure.
   e. nuclear reactors and radioactive wastes.

E 21. The usual, naturally-occurring mechanism that prevents development of cancer is the action of
 * a. cytotoxic T cells.
   b. oncogenes.
   c. radiation.
   d. chemotherapy.
   e. carcinogens.

DIAGNOSING CANCER

E 22. Detection of cancer by blood testing depends on the presence of __?__ in the blood.
   a. malignant cells
   b. mutant DNA
 * c. tumor markers
   d. carcinogens
   e. cytotoxic T cells

E 23. Monoclonal antibodies are useful in detecting cancer because they home in on
 * a. tumor antigens.
   b. hormones released by the tumors.
   c. interleukins.
   d. oncogenes.
   e. carcinogens.

E 24. The most definitive tool in cancer detection is
a. monoclonal antibody.
b. medical imaging.
c. screening.
* d. biopsy.
e. computerized tomography.

TREATING AND PREVENTING CANCER

E 25. Which of the following is a goal of chemotherapy?
* a. disruption of the mitotic spindle
b. alteration of the plasma membrane
c. increasing the secretion of interleukins
d. stimulating production of cytotoxic T cells
e. turning on tumor suppressor genes

M 26. Which of the following cancer treatments is LEAST likely to kill healthy cells?
a. chemotherapy
b. adjuvant therapy
* c. immunotherapy
d. radiation
e. fluorouracil

E 27. All of the following statements are ways to reduce cancer risk except one. Which is it?
a. Restrict alcohol intake.
* b. Begin estrogen replacement therapy after menopause.
c. Maintain desirable weight.
d. Avoid tobacco.
e. Avoid direct sun exposure.

SOME MAJOR TYPES OF CANCER

E 28. Cancers of the epithelium are called
a. sarcomas.
b. leukemias.
* c. carcinomas.
d. lymphomas.
e. adenocarcinomas

E 29. Cancers of connective tissues such as bone are called
* a. sarcomas.
b. leukemias.
c. carcinomas.
d. lymphomas.
e. adenocarcinomas

E 30. Cancers of blood-forming regions are called
a. sarcomas.
* b. leukemias.
c. carcinomas.
d. lymphomas.
e. adenocarcinomas

E 31. Which of the following cancers causes the greatest number of deaths in both men and women?
a. colon
b. reproductive
* c. lung
d. skin
e. blood

CANCERS OF THE BREAST AND REPRODUCTIVE SYSTEM

E 32. Which of the following is NOT associated in some way with the topic of breast cancer?
a. mammography
b. high cure rate
* c. Pap smear
d. lumpectomy
e. mastectomy

M 33. Which of the following cancers is often lethal because of difficulties in detection until metastasis has occurred?
* a. ovarian
b. breast
c. uterus
d. prostate
e. skin

E 34. The PSA test is one of the newest methods for detecting cancers of the
a. uterus.
b. breast.
c. colon.
* d. prostate.
e. lung

A SURVEY OF OTHER COMMON CANCERS

E 35. The overwhelming risk factor for cancers of the mouth and lungs is
a. a high fat diet.
* b. use of tobacco products.
c. exposure to radiation.
d. hormonal imbalances in puberty.
e. excessive tanning.

E   36. Hodgkin's disease is cancer of
         a. skin.
         b. pancreas.
         c. urinary tract.
      * d. lymph organs.
         e. breast.

E   37. Malignant melanoma is highly dangerous
         because
         a. it is so difficult to detect.
         b. there are virtually no effective
            treatments.
      * c. it metastasizes so aggressively.
         d. it is contagious.

Matching Questions

D   38. Matching. Choose the best matching
         element.
         1 ____   biopsy
         2 ____   carcinogen
         3 ____   metastasis
         4 ____   oncogene
         5 ____   tumor marker
         6 ____   tumor suppressor gene
            A. encodes proteins that keep cell
               growth and division within normal
               bounds
            B. DNA segment that can induce cancer
               in normal cell
            C. migration of cancer cells
            D. substances produced in response to
               cancer
            E. removal of suspect tissue for
               microscopic evaluation
            F. cancer-causing substance

Answers: 1. E   2. F   3. C
             4. B   5. D   6. A

Classification Questions

Answer questions 39–43 in reference to the five types of
cancer listed below:
         a. sarcoma
         b. carcinoma
         c. adenocarcinoma
         d. lymphoma
         e. leukemia

D   39. This a cancer of epithelium, including cells
         of the skin and internal linings.

D   40. This cancer begins in the body or ducts of a
         gland.

M   41. This is a cancer of the blood-forming
         regions of the bone marrow.

M   42. This is a cancer of connective tissue such as
         bone.

M   43. This is a cancer of the lymphoid tissues.

Answers: 39. b   40. c   41. e
             42. a   43. d

Selecting The Exception

M   44. Four of the five answers listed below are
         forms of immune therapy for cancer. Select
         the exception.
         a. interferon
      * b. fluorouracil
         c. interleukins
         d. cytotoxic T cells
         e. vaccines

E   45. Four of the five answers listed below are
         methods of cancer treatment. Select the
         exception.
         a. chemotherapy
         b. surgery
         c. lymphokine-activated killer cells
         d. adjuvant therapy
      * e. magnetic resonance imaging

M   46. Four of the five answers listed below are
         methods of cancer detection. Select the
         exception.
         a. biopsy
         b. magnetic resonance imaging
         c. monoclonal antibodies
      * d. interleukins
         e. screening

D **47.** Four of the five answers listed below are
          descriptions of cancer cells. Select the
          exception.
  *   a. abnormally-shaped nucleus
      b. decline in ability to adhere to substrates
      c. changes in the plasma membrane
      d. abnormal growth and division
      e. cytoplasm shrinks and becomes
          disorganized

E **48.** Four of the five answers listed below are
          carcinogens. Select the exception.
  *   a. egg white
      b. asbestos
      c. radiation with x-ray
      d. components in cigarette smoke
      e. ultraviolet radiation

CHAPTER 23
PRINCIPLES OF EVOLUTION

Multiple-Choice Questions

UNITY AND DIVERSITY

E **1.** Which of the following includes all the others?
   a. family
   * b. phylum
   c. species
   d. class
   e. order

E **2.** Which includes all related genera?
   * a. family
   b. phylum
   c. species
   d. class
   e. order

M **3.** Phylogeny refers to what aspects of individuals?
   a. morphological traits
   * b. evolutionary relationships
   c. physiological characteristics
   d. behavioral features
   e. all of these

D **4.** The assigning of scientific names is called
   a. gradualism
   b. convergence
   c. classification
   * d. taxonomy
   e. phylogeny

E **5.** "Taxa" refers to
   a. species.
   b. genus.
   c. family.
   d. phylum.
   * e. all of these

M **6.** Which of the following groups represents the most closely related organisms?
   a. kingdoms
   * b. species
   c. orders
   d. genera
   e. taxa

M **7.** Organisms "X" and "Y" are suspected to be the same species. Which of the following will provide the ultimate proof?
   * a. interbreeding
   b. anatomy
   c. physiology
   d. ecology
   e. behavior

M **8.** Which of the following is the least inclusive category?
   a. family
   b. order
   * c. species
   d. kingdom
   e. genus

M **9.** The only taxonomic category in which evolution can occur is the
   a. genus.
   * b. species.
   c. kingdom.
   d. family.
   e. class.

A LITTLE EVOLUTIONARY HISTORY

E **10.** Charles Darwin
   a. lived in England.
   b. studied to become a clergyman.
   c. followed his avocation as a naturalist.
   * d. all of these
   e. none of these

E **11.** Darwin's mentor, who obtained Darwin's position on H.M.S. *Beagle*, was
   a. Alfred Russell Wallace.
   * b. John Henslow.
   c. Jean-Baptiste Lamarck.
   d. Georges Cuvier.
   e. Charles Lyell.

D   **12.** After his return to England, Darwin pondered which of the following questions most heavily?
*   a. What could explain the great numbers and variety of species?
    b. Does the fossil evidence support uniformitarianism?
    c. Are the extinct and living armadillos the same species?
    d. Did Galápagos finches have a common mainland ancestor?
    e. Will natural selection work in England?

E   **13.** Thomas Malthus proposed that
    a. the food supply multiplied faster than the population.
*   b. the population multiplied faster than the food supply.
    c. the food supply and population multiplied at the same rate.
    d. artificial selection was the key to evolution.
    e. natural selection was the key to evolution.

SOME BASIC PRINCIPLES OF EVOLUTION

M   **14.** Microevolution is the result of
    a. chance variation.
    b. change in gene frequency.
    c. mutation.
    d. natural selection.
*   e. all of these

E   **15.** The word *evolution* as used in biology literally means
    a. natural selection.
    b. genetic drift.
    c. divergence.
*   d. change.
    e. mutation.

M   **16.** Which is a group of individuals of the same species for which there are no restrictions to random mating among its members?
    a. individual
    b. species
*   c. population
    d. polyploid
    e. all of these

E   **17.** Which of the following evolve?
*   a. populations
    b. genera
    c. kingdoms
    d. populations and genera
    e. genera and kingdoms

D   **18.** Natural selection operates to produce changes in
    a. individuals.
*   b. populations.
    c. races.
    d. phyla.
    e. animals only.

E   **19.** New variations of genes may be produced by
    a. immigration.
    b. mutation.
    c. crossing over.
    d. sexual reproduction.
*   e. all of these

M   **20.** Members of a population cannot have which of the following in common?
    a. phenotype
    b. morphological traits
*   c. genotype
    d. physiological traits
    e. behavioral traits

D   **21.** Only identical twins have the same
*   a. genotype.
    b. phenotype.
    c. traits.
    d. genotype and phenotype.
    e. genotype, phenotype, and traits.

PROCESSES OF MICROEVOLUTION

M   **22.** Introduction of previously nonexistent genes into a population may be accomplished by
    a. nonrandom mating.
*   b. mutation.
    c. sexual recombination.
    d. the founder effect.

E   **23.** New alleles arise by
*   a. mutation.
    b. migration.
    c. genetic drift.
    d. random mating.
    e. independent assortment.

M 24. New alleles that appear by mutation
   a. are inherently disadvantageous to their bearers.
   b. are seldom advantageous or disadvantageous in themselves.
   c. either have or lack survival value only in the context of their environment.
   * d. are seldom advantageous or disadvantageous in themselves and either have or lack survival value only in the context of their environment.
   e. are inherently disadvantageous to their bearers and are seldom advantageous or disadvantageous in themselves.

D 25. One part of Darwin's theory is that individuals with certain traits have an increased competitive edge. The source of these traits is
   a. adaptation to the stress.
   b. development over a lifetime.
   * c. inheritance from birth.
   d. mutation after birth.
   e. all of these

M 26. The operation of natural selection depends upon the fact that
   a. the strong always survive, whereas the weak always die.
   * b. some individuals have a better chance to produce more offspring.
   c. mutations are always harmful.
   d. acquired characteristics are inherited.
   e. reproduction of all members of a species is virtually the same.

M 27. What accounts for the fact that polydactylism is prevalent and Tay-Sachs disease virtually absent in one human population in the United States while Tay-Sachs disease is prevalent and polydactylism virtually absent in another?
   a. Natural selection has promoted these differences since humans live in many different environments.
   b. Mutation rates differ between different loci.
   * c. There is little gene flow between the two populations.
   d. The populations are small, and therefore genetic drift is a major factor in the determination of allele frequencies.

M 28. The introduction of a small population onto an island that results in a limited gene pool for a population best describes
   a. the Hardy-Weinberg law.
   b. genetic drift.
   c. the bottleneck effect.
   * d. the founder principle.
   e. the effect of genetic isolation.

E 29. The influence of genetic drift on allele frequencies increases as
   a. gene flow increases.
   * b. population size decreases.
   c. mutation rate decreases.
   d. the number of heterozygous loci increases.

M 30. The evolutionary force that operates primarily through chance is
   a. natural selection.
   * b. genetic drift.
   c. isolation.
   d. mating preference.

D 31. Although there are as many starlings in North America as there are in Europe, genetic variability in the North American population is reduced relative to that in Europe because
   a. there are more environments in Europe.
   * b. the North American population is derived from a small founder population.
   c. there is more gene flow in Europe.
   d. there is less mutation in North America.

E 32. All of the following are components of Darwin's principle of natural selection EXCEPT
   * a. new alleles are constantly produced through mutation.
   b. populations exhibit great variation.
   c. organisms produce more offspring than can be sustained by the environment.
   d. over time, adaptive phenotypes increase in frequency within a population.

E   33. Natural selection is best defined as
  a. mutation in various species.
  b. independent assortment of genes during meiosis.
  * c. the difference in survival and reproduction in a population.
  d. changes in chromosome number or structure.
  e. the adaptation of varying individuals to changes in the environment.

E   34. The trend in which organisms come to have characteristics that suit them to conditions in a particular environment is called
  a. genetic drift.
  * b. adaptation.
  c. mutation.
  d. isolation.
  e. divergence.

E   35. An insect that exhibits resistance to a pesticide
  a. developed the resistance in response to the pesticide.
  b. mutated when exposed to the pesticide.
  * c. inherited genes that made it resistant to the pesticide.
  d. none of these

M   36. When DDT was first introduced, insects were very susceptible to it. The development of resistance to DDT by insects was the result of
  a. special creation.
  * b. natural selection of forms that expressed genes for resistance.
  c. the high biotic potential of insects.
  d. a naturally occurring example of inheritance of acquired characteristics.
  e. mutation induced by DDT.

E   37. A species is composed of
  a. related organisms.
  b. a group of reproductive females.
  * c. populations that have the potential to interbreed and produce fertile offspring.
  d. organisms located in the same habitat.
  e. all males and females in the same geographical range with the same ecological requirements.

E   38. The term reproductive isolation mechanism refers to
  a. specific areas where males compete or display for females.
  b. the process by which sexual selection evolves within a population.
  * c. a blockage of gene flow between populations.
  d. the inability of a species to continue reproduction.

D   39. Incompatibilities between the developing embryo and the maternal organism that cause the embryo to abort spontaneously may prevent individuals of different populations from producing fertile offspring. Such differences may be which of the following?
  * a. isolating mechanisms
  b. allele frequencies
  c. mutations
  d. founder effects
  e. gene flow

M   40. Two individuals are members of the same species if they
  a. possess the same number of chromosomes.
  b. breed at the same time.
  c. are phenotypically indistinguishable.
  * d. can mate and produce fertile offspring.

E   41. Complete reproductive isolation is evidence that what has occurred?
  a. extinction
  * b. speciation
  c. polyploidy
  d. hybridization
  e. gene flow

D   42. Divergence may lead to
  a. genetic drift.
  * b. speciation.
  c. balanced polymorphism.
  d. gene flow.
  e. genetic equilibrium.

E   43. The punctuated equilibrium model of evolutionary change proposes that most morphological change occurs
  a. gradually within a species.
  b. rapidly within a species.
  c. gradually during speciation.
  * d. rapidly during speciation.

M **44.** According to the punctuated equilibrium model of speciation, a tree of descent would be characterized by
* a. vertical lines with horizontal branchings.
  b. vertical lines with branchings at narrow angles.
  c. broad-angled lines with horizontal branches.
  d. broad-angled lines with branching at narrow angles.

LOOKING AT FOSSILS AND BIOGEOGRAPHY

M **45.** Macroevolution refers to changes in all but which one of the following?
  a. phyla
  b. classes
* c. species
  d. genera
  e. divisions

E **46.** The study of the worldwide distribution of plants and animals is called
  a. paleontology.
  b. microevolution.
  c. comparative morphology.
* d. biogeography.
  e. phylogeny.

D **47.** Plate tectonic theory is based on
  a. a thermal convection model, in which cool material in the earth's mantle rises and spreads laterally beneath the crustal plates.
  b. the idea that the earth's crust is fragmented into rigid crusts that are sinking slowly beneath crustal plates.
  c. the idea that coacervate formation causes continents to drift apart slowly on their crustal plates.
* d. observations that the sea floor is slowly spreading away from oceanic ridges due to thermal convection in the mantle.
  e. all of these

E **48.** Fossils found in the lowest geological strata are generally the most
  a. advanced.
  b. complex.
* c. primitive.
  d. widespread.
  e. specialized.

E **49.** The fossil record is incomplete because
  a. very few organisms were preserved as fossils.
  b. organisms tend to decay before becoming a fossil.
  c. animals with hard parts are preserved more easily.
  d. geological processes may destroy fossils.
* e. all of these

E **50.** Fossils would include
  a. skeletons.
  b. shells.
  c. seeds.
  d. tracks.
* e. all of these

M **51.** Which of the following organisms would you expect to find preserved as a fossil?
  a. a jellyfish
* b. a shelled arthropod such as a trilobite
  c. an earthworm
  d. a nematode
  e. a protozoan

M **52.** Which of the following statements is false?
  a. Many fossils have not been discovered, whereas others may have been destroyed.
  b. Some types of organisms are more likely to be preserved than others.
  c. Some environments are more conducive to preserving.
  d. "Many fossils have not been discovered, whereas others may have been destroyed" and "Some types of organisms are more likely to be preserved than others."
* e. "Many fossils have not been discovered, whereas others may have been destroyed," "Some types of organisms are more likely to be preserved than others," and "Some environments are more conducive to preserving."

M **53.** Which of the following habitats is most likely to be rich in fossils?
  a. eroding hillsides
  b. deserts
  c. polar ice caps
* d. bed of former shallow sea
  e. rocky plateau

COMPARING THE FORM AND DEVELOPMENT OF BODY PARTS

M 54. The study of comparative morphology has revealed the conservative nature of the genes responsible for
    a. food procurement.
    b. reproductive behavior.
   * c. embryonic development.
    d. size.
    e. mating behavior.

M 55. The convergence in external morphology of sharks, penguins, and porpoises is attributed to
    a. reduced genetic variability in these groups.
   * b. selection pressures that are common to these groups.
    c. reproductive isolation of these groups.
    d. identical genes in all three groups.

M 56. Which serve as examples of convergence?
   * a. penguins and porpoises
    b. panthers and tigers
    c. apes and monkeys
    d. sharks, skates, and rays
    e. mice, rats, and gerbils

M 57. Phylogenetic relationships, when determined solely by the study of comparative morphology, may be incorrect due to
    a. morphological divergence.
   * b. convergence.
    c. adaptive radiation.
    d. extinction.

D 58. The wings of a bird and the wings of a butterfly are _____ and show morphological _____.
    a. homologous; convergence
   * b. analogous; convergence
    c. homologous; divergence
    d. analogous; divergence

COMPARING BIOCHEMISTRY

M 59. All of the following are useful indicators of phylogenetic relatedness EXCEPT
    a. base sequences in DNA.
    b. amino acid sequences in a protein.
   * c. similar ecological requirements.
    d. similar embryonic development.

M 60. Which of the following has been used to measure more precisely the relatedness of primates?
   * a. cytochrome c
    b. blood type
    c. family trees
    d. convergence
    e. fossils

D 61. Comparisons of protein similarity between species can reveal the degree of genetic kinship because
    a. the number of protein variations is limited.
   * b. specific amino acids are dictated by known nucleotide sequences.
    c. gel electrophoresis converts proteins to nucleotides.
    d. protein can be hybridized with DNA.
    e. DNA is made by directions stored in proteins.

D 62. A molecular clock
    a. can tell how old a particular individual organism is.
    b. is used to date fossils.
    c. is calibrated by monitoring radioactive decay.
   * d. can give some indication of the time of divergence of species from one another.
    e. is based on astronomy data.

EVOLUTIONARY TREES AND THEIR BRANCHINGS

E 63. The acquisition of a key evolutionary innovation by a species gives evidence for the concept of
    a. uniformitarianism.
    b. gradualism.
    c. convergence.
   * d. adaptive radiation.
    e. special creation.

E 64. Background extinction is a measure of
    a. the rate of species turnover at the end of geological eras.
    b. the number of species that suffer extinction at the beginning of geological eras.
   * c. the steady rate of species turnover within a lineage throughout most of their evolutionary history.
    d. the lowest rate of species turnover within a lineage observed within a geological era.

M 65. Explanations for mass extinction include all of the following EXCEPT
  a. collisions between the earth and other bodies in the solar system.
  b. continental movements.
  * c. adaptive radiation of new predator species in many lineages.
  d. alterations in sea level.

M 66. It is thought that the earliest tools were employed by hominids to
  a. assist in locomotion.
  b. provide protection.
  * c. facilitate the processing of food.
  d. ward off predators.

E 67. The oldest "manufactured" tools have been found in
  a. North America.
  b. Eurasia.
  * c. Africa.
  d. Australia.

M 68. The geographical distribution of hominids changed dramatically about 2 million years ago due to the migrations of
  a. *Australopithecus robustus.*
  b. *Australopithecus boisei.*
  * c. *Homo erectus.*
  d. *Homo sapiens.*

M 69. The first toolmakers were
  a. *Australopithecus africanus.*
  b. *Australopithecus robustus.*
  c. *Australopithecus boisei.*
  * d. *Homo habilis.*
  e. *Homo erectus.*

E 70. The oldest hominid fossils have been found in
  a. North America.
  b. Eurasia.
  * c. Africa.
  d. Australia.

TRENDS IN HUMAN EVOLUTION

E 71. Dentition refers to what aspects of teeth?
  a. type
  b. number
  c. size
  d. type and number.
  * e. type, number, and size.

M 72. The study of dentition tells the researcher what about an animal?
  a. its diet
  b. its life-style
  c. its intelligence
  * d. its diet and life-style
  e. its diet, life-style, and intelligence

E 73. Bipedalism is most highly developed in
  a. hominoids.
  b. apes.
  * c. humans.
  d. monkeys.
  e. prosimians.

M 74. In the course of the evolution of existing primate groups, there has been a general decrease in
  * a. number of offspring produced by a female.
  b. body size.
  c. life span.
  d. duration of infant dependency.

E 75. All but which factor were important evolutionary adaptations in primates?
  a. enhanced stereoscopic vision
  b. upright position
  c. an opposable thumb
  * d. the development of a restricted or specialized diet
  e. brain expansion and elaboration

M 76. Which feature is NOT characteristic of the evolutionary trends in primates?
  a. longer life span
  b. longer gestation period
  c. longer infant dependency
  d. longer periods between pregnancies
  * e. larger litters

M 77. Behavioral trends in primate evolution include
  a. longer life spans.
  b. longer learning period and dependence on parents.
  c. lower reproductive rate.
  d. longer periods between pregnancies.
  * e. all of these

M 78. In the evolution of the arboreal primates, which of the following features would NOT be an important evolutionary advancement?
    a. opposable thumbs
    b. enhanced daytime vision
  * c. elongated snout with well-developed sense of smell
    d. the ability to see in color
    e. a brain with a well-developed ability to judge distances and with the ability to quickly compensate for misjudgments

M 79. The most recent level of evolution in primates is considered to be in
    a. brain expansion.
  * b. behavior and culture.
    c. dentition.
    d. hand grip.
    e. daytime vision.

M 80. The primates first arose about how many million years ago?
    a. 75
  * b. 60
    c. 50
    d. 40
    e. 30

M 81. Primitive primates generally live
  * a. in tropical and subtropical forest canopies.
    b. in temperate savanna and grassland habitats.
    c. near rivers, lakes, and streams in the East African Rift Valley.
    d. in caves with abundant supplies of insects.
    e. all of these

M 82. The first known primates were characterized as
  * a. arboreal and nocturnal.
    b. ground-dwelling and nocturnal.
    c. arboreal and diurnal.
    d. ground-dwelling and diurnal.

EVOLUTION AND EARTH HISTORY

E 83. The primitive earth's atmosphere did NOT contain
    a. water vapor.
    b. free nitrogen.
    c. free hydrogen.
  * d. free oxygen.
    e. inert gases.

E 84. Many of the organic compounds essential for life, such as amino acids and nucleotides, could NOT assemble spontaneously in the presence of
    a. hydrogen.
  * b. free oxygen.
    c. carbon dioxide.
    d. nitrogen.
    e. argon.

E 85. Organic compounds break down spontaneously in the presence of _____; hence, life probably never would have emerged if the ancient atmosphere had been the same as the present one.
    a. carbon dioxide
    b. hydrogen
  * c. oxygen
    d. nitrogen
    e. silica

THE ORIGIN OF LIFE

E 86. Experiments like those first performed by Stanley Miller in 1953 demonstrated that
    a. DNA forms readily and reproduces itself.
  * b. many of the lipids, carbohydrates, proteins, and nucleotides required for life can form under abiotic conditions.
    c. complete, functioning prokaryotic cells are formed after approximately three months.
    d. a lipid-protein film will eventually be formed by thermal convection.
    e. all of these

E 87. Which of the following was NOT included in Miller's reaction chamber, which contained substances intended to duplicate the atmosphere of ancient earth?
  * a. carbon dioxide
    b. methane
    c. ammonia
    d. water vapor
    e. both methane and ammonia

E 88. Who demonstrated the possibility of producing organic compounds from gases and water if the mixture is bombarded with a continuous spark discharge?
  * a. Miller
    b. Starr
    c. Thompsen
    d. Pauling

E 89. The Miller experiment, a study of the early synthesis of organic compounds included all of the following molecules EXCEPT
   a. methane.
   b. ammonia.
   c. water.
* d. oxygen.

M 90. Protein synthesis on the primordial earth may have been catalyzed by _____ before the evolution of enzymes.
   a. DNA
   b. carbohydrates
   c. amino acids
* d. RNA
   e. lightning

M 91. The formation of polypeptide chains under abiotic conditions was important because they served as
   a. a supply of structural units.
   b. enzymes to catalyze reactions.
   c. subunits in the formation of DNA.
   d. subunits in the formation of RNA.
* e. both a supply of structural units and enzymes to catalyze reactions.

M 92. What step occurred first in the evolution of life?
   a. formation of lipid spheres
   b. formation of protein-RNA systems
   c. formation of membrane-bound protocells
* d. spontaneous formation of lipids, proteins, carbohydrates, and nucleotides under abiotic conditions
   e. formation of ATP

M 93. Which step in the evolution of life is the most complex and occurred last?
   a. formation of lipid spheres
   b. formation of protein-RNA systems
* c. formation of membrane-bound protocells
   d. spontaneous formation of lipids, proteins, carbohydrates, and nucleotides under abiotic conditions
   e. formation of ATP

D 94. Contemporary hypotheses concerned with the origin of life focus on what two characteristics of living systems?
   a. energy conversion and development of a nucleus
   b. self-replication and utilization of oxygen
* c. plasma membranes and self-replication
   d. growth and transcription

M 95. Sidney Fox found that if heated protein chains were allowed to cool in water they would
   a. form nitrogen, which would escape as a gas.
   b. form proteinoids.
* c. form small, stable spheres or microspheres.
   d. clot and form a complex latticework frame for chemical reactions.
   e. break down into the original amino acids that the protein chain was made from.

Classification Questions

Answer questions 96-99 in reference to the four evolutionary processes listed below:

   a. mutation
   b. gene flow
   c. genetic drift
   d. natural selection

M 96. This is most likely to lead to the loss of genetic variation in a small population.

E 97. This process produces new genetic variation within a species.

M 98. This process can rapidly offset the effects of genetic isolation when two populations come into secondary contact.

D 99. The reduced contribution of one phenotype in comparison to another to the next generation is an example of this.

Answers: 96. c   97. a   98. b
              99. d

Answer questions 100-103 in reference to the five taxonomic categories listed below:

    a. genus
    b. species
    c. order
    d. family
    e. phyla

E   **100.** This category is not included in any of the other listed categories.

E   **101.** This category is included in each of the other categories.

M   **102.** The term *Hominidae* is an example of this category.

E   **103.** This category denotes the taxonomic category of *Homo*, which describes humans.

Answers: 100. e   101. b   102. d
            103. a

Answer questions 104-108 in reference to the four terms listed below:

    a. gradualism
    b. convergence
    c. punctuation
    d. divergence

M   **104.** The porpoise and the penguin are examples of this.

M   **105.** The accumulation of allelic differences between two species over a period of 2 million years is an example of this.

M   **106.** The rapid divergence of two reproductively isolated groups following speciation is an example of this.

D   **107.** Wings of pterosaurs, birds, and bats are examples of this.

D   **108.** Morphological change that occurs within species by changes in allele frequencies is an example of this.

Answers: 104. b   105. a   106. c
            107. b   108. a

Selecting the Exception

M   **109.** Four of the five answers listed below are sources of variation in a population. Select the exception.
    a. mutation
    b. sexual reproduction
    c. crossing over
    d. independent assortment
*   e. law of dominance

E   **110.** Four of the five answers listed below are characteristics of mutations. Select the exception.
*   a. predictable
    b. lethal or beneficial
    c. random
    d. effects depend upon environment
    e. heritable

M   **111.** Four of the five answers listed below are portions of the theory of natural selection. Select the exception.
    a. variation is heritable
    b. heritable traits vary in adaptability
    c. more organisms are produced than can survive
*   d. the largest and strongest always contribute more genes to the next generation
    e. natural selection is the result of differential reproduction

D   **112.** Four of the five answers listed below are habitats favoring fossil preservation. Select the exception.
*   a. deserts
    b. swamp
    c. tar pits
    d. seafloor
    e. caves

M   **113.** Four of the five answers listed below are components of the mixture used in Miller's experiment. Select the exception.
    a. hydrogen
*   b. oxygen
    c. methane
    d. ammonia
    e. water

E   **114.** Four of the five answers below are taxonomic categories. Select the exception.
    a. species
    b. class
*   c. taxon
    d. order
    e. phylum

CHAPTER 24
ECOSYSTEMS

Multiple-Choice Questions

INTRODUCTION TO PRINCIPLES OF ECOLOGY

E **1.** Which is a habitat?
   a. predator
 * b. intestinal tract
   c. parasite
   d. producer
   e. decomposer

M **2.** All of the populations of different species that occupy and are adapted to a given area are referred to by which term?
   a. biosphere
 * b. community
   c. ecosystem
   d. niche
   e. habitat

M **3.** What term denotes the range of all factors that influence whether a species can obtain resources essential for survival and reproduction?
   a. habitat
 * b. niche
   c. carrying capacity
   d. ecosystem
   e. community

M **4.** Which statement about ecosystems is false?
   a. The rate of energy flow depends on the ratio of producers to consumers.
   b. The requirements of an ecosystem change with age.
   c. The larger the ecosystem, the more flexible it is.
 * d. The smaller the ecosystem, the more stable it is.
   e. The more efficient the producers are, the more energy must be put in and the more energy is available for the next trophic level.

M **5.** A network of interactions that involve the cycling of materials and the flow of energy between a community and its physical environment is which of the following?
   a. population
   b. community
 * c. ecosystem
   d. biosphere
   e. species

E **6.** The regions of the earth's crust, waters, and atmosphere in which organisms can live is defined as the
   a. ecosystem.
   b. community.
 * c. biosphere.
   d. habitat.
   e. niche.

E **7.** The study of the interactions of organisms and their physical environment is
   a. biology.
 * b. ecology.
   c. geology.
   d. paleontology.
   e. physiology.

M **8.** A community differs from an ecosystem in that the former does NOT include
   a. unicellular organisms.
   b. decomposers.
 * c. abiotic (nonliving) factors.
   d. unicellular organisms and decomposers.
   e. unicellular organisms, decomposers, and abiotic (nonliving) factors.

M **9.** Secondary succession is likely to occur
   a. in a deciduous forest.
   b. on an eroded, bare hillside.
   c. in an abandoned field.
   d. in a deciduous forest and on an eroded, bare hillside.
 * e. in a deciduous forest, on an eroded, bare hillside or in an abandoned field.

M **10.** Secondary succession occurs
 * a. after a fire.
   b. on a new sand dune.
   c. on bare rock.
   d. immediately after the formation of a man-made lake.

M 11. The plants and animals now present on acreage from which the trees were removed ten years earlier represent
    a. primary succession.
    b. a climax forest.
    c. pioneer species.
    * d. secondary succession.
    e. species introductions.

D 12. Farmland that is under regular and continued tillage will not
    a. undergo succession.
    * b. produce a climax community.
    c. experience competition.
    d. suffer from the effects of disturbance.
    e. develop species diversity.

M 13. The climax community
    a. is formed by species with the greatest range of environmental tolerance.
    b. is the most common community found in an area.
    c. changes over time.
    * d. is well adapted to the climate and persists until the climate changes.

M 14. Which of the following would be more likely to affect an animal's habitat than its niche?
    * a. rainfall
    b. prey abundance
    c. predators
    d. defense mechanisms

M 15. When Shakespeare wrote about the world as a stage and each of us being players, he was unknowingly referring to the biological concept of
    a. succession.
    * b. the niche.
    c. different habitats.
    d. feeding levels.
    e. interspecific competition.

E 16. Niche refers to the
    a. home range of an animal.
    b. preferred habitat for an organism.
    * c. functional role of a species in a community.
    d. territory occupied by a species.

THE NATURE OF ECOSYSTEMS

E 17. In a natural community, the primary consumers are
    * a. herbivores.
    b. carnivores.
    c. scavengers.
    d. decomposers.
    e. all of these

E 18. Which is usually a primary carnivore?
    a. chicken
    b. cow
    c. rabbit
    * d. wolf
    e. squirrel

M 19. Which is a primary consumer?
    * a. cow
    b. dog
    c. hawk
    d. fox
    e. snake

M 20. Which of the following are NOT heterotrophs?
    a. primary carnivores
    b. herbivores
    c. detritivores
    d. decomposers
    * e. All of these are heterotrophs.

D 21. Which of the following is NOT dependent on the others as a food supply?
    a. carnivores
    b. herbivores
    * c. producers
    d. detritivores
    e. decomposers

M 22. The primary consumer is also
    a. the second link in a food chain.
    b. a herbivore.
    c. an animal.
    d. a herbivore and an animal.
    * e. the second link in a food chain plus being a herbivore and an animal.

D 23. A secondary consumer could eat
    a. only herbivores.
    b. only primary producers.
    c. primary carnivores.
    * d. anything "below" it in the food web.

D 24. Food chains rarely have more than three levels of consumers because
    a. the animals are too large to search for prey.
    b. the growing season of plants is not long enough.
    c. pyramids do not go that high.
  * d. the amount of energy still available is too small.

D 25. After a shipwreck you are alone on a deserted island with a pair of pigs and a large supply of corn. Your best strategy would be to
    a. feed the corn to the pigs and feed upon their offspring.
  * b. kill the pigs immediately and then eat the corn.
    c. share the corn with the pigs and then eat the pigs when the corn is gone.

M 26. Detritivores are
    a. bacteria.
    b. plants.
    c. fungi.
  * d. animals.
    e. bacteria and fungi.

E 27. Herbivores represent the
  * a. primary consumers.
    b. secondary consumers.
    c. tertiary consumers.
    d. primary producers.
    e. secondary producers.

ENERGY FLOW THROUGH ECOSYSTEMS

E 28. The ultimate source of all energy in a terrestrial ecosystem is
    a. the organic matter in all the organisms of the ecosystem.
    b. water.
  * c. sunlight.
    d. carbon dioxide.

M 29. Primary carnivores are
    a. tertiary consumers in the third trophic level.
  * b. secondary consumers in the third trophic level.
    c. secondary consumers in the second trophic level.
    d. tertiary consumers in the fourth trophic level.

M 30. Which cannot be placed in a single trophic level?
    a. oak tree
    b. zebra
  * c. mushroom
    d. rabbit

M 31. All living organisms are dependent upon plants because
    a. plants produce oxygen as a by-product of photosynthesis.
  * b. as producers, they form the base of food chains.
    c. they function to prevent erosion and reduce desertification.
    d. as they remove carbon dioxide from the atmosphere, they reduce the problems generated by the greenhouse effect.

D 32. Which of the following combinations of organisms could be expected to survive in isolation from other forms of life available?
  * a. producers and decomposers
    b. producers and carnivores
    c. carnivores and decomposers
    d. herbivores, carnivores, and decomposers

M 33. Wastes would accumulate and most nutrients would stop cycling if the _____ in the ecosystem died.
    a. protozoans and protistans
  * b. bacteria and fungi
    c. flatworms, roundworms, and earthworms
    d. insects

E 34. Which of the following is the correct word meaning "to feed"?
    a. tropic
  * b. trophic
    c. topic
    d. tophic
    e. tropical

D 35. Most of the energy available to a primary consumer
  * a. will be used up in various biological activities.
    b. will be converted into biomass.
    c. is obtained directly from solar energy.
    d. will be passed on to the animal that feeds upon it.

D 36. Net primary productivity is the
a. rate of photosynthesis.
b. rate of energy flow.
c. amount of energy stored in the ecosystem.
d. amount of energy utilized.
* e. amount of energy stored in the plant tissue in excess of that used by autotrophs in respiration.

M 37. Consumption of _____ is _____ in plants than in animals.
* a. oxygen; less
b. carbon dioxide; less
c. oxygen; greater
d. carbohydrate; greater
e. carbon monoxide; greater

D 38. The amount of energy that flows through a detrital food web is _____ that which flows through a grazing web.
a. the same as
* b. greater than
c. less than
d. the sum of
e. the difference of

M 39. The difference between gross primary productivity and net primary productivity is
a. the amount of sunlight reflected by plants.
b. the rate of photosynthesis of autotrophs.
* c. the rate of respiration of autotrophs.
d. the rate of herbivorous consumption of autotrophs.

M 40. The simple food chain: grass >>> zebra >>> lion, provides a good example of a pyramid of
a. energy.
b. height.
c. biomass.
d. energy and height.
* e. energy and biomass.

M 41. Decomposers perform their recycling efforts on organisms
a. at the end of a food chain.
b. on the top of a pyramid.
c. that are producers.
d. that are consumers.
* e. all of these

E 42. Decomposers
a. are able to enter a food chain at any trophic level.
b. are the most numerous organisms in an ecosystem.
c. include bacteria and fungi.
* d. all of these

M 43. Detritus specifically includes
a. organic wastes.
b. toxic materials.
c. dead and partially decayed material.
d. living bacteria and fungi.
* e. organic wastes plus dead and partially decayed material.

E 44. At the bottom or base of a pyramid of energy are the
* a. primary producers.
b. secondary producers.
c. primary consumers.
d. secondary consumers.
e. tertiary consumers.

D 45. The pyramid of energy is
a. a demonstration of the first law of thermodynamics.
* b. a result of the decline in the energy available as energy travels through the trophic levels.
c. fundamentally different from the pyramid of biomass and the pyramid of numbers.
d. just one of the manifestations of competition.

E 46. At the top of a pyramid of biomass are the
a. primary producers.
b. secondary producers.
c. primary consumers.
d. secondary consumers.
* e. tertiary consumers.

M 47. The biomass of a community is the weight of the
a. material decomposed in a year.
b. producers.
* c. living organisms.
d. consumers.
e. decomposers.

SCIENCE COMES TO LIFE: ENERGY FLOW AT SILVER SPRINGS

M   48. Which of the following is NOT true of ecosystems?
   a. Although they may include many different species, many features of ecosystem structure and function are alike.
   b. Autotrophs secure energy and nutrients that are then used by heterotrophs.
* c. Energy cycles and minerals flow through ecosystems.
   d. Many different niches are represented in most ecosystems.
   e. Ecosystems are characterized by relatively few trophic levels.

E   49. Most of the energy within an ecosystem is lost
   a. when organisms disperse.
   b. when organisms die.
* c. as a result of metabolism.
   d. by organisms at the top of the food web.

E   50. Of the energy that enters one trophic level, approximately what percent (average) becomes available for the next trophic level?
   a. 100
* b. 10
   c. 1
   d. 0.1
   e. 0.01

M   51. Which of the following statements is false?
   a. Heat loss represents a one-way loss of energy from an ecosystem.
* b. Organisms in the food chain use all the energy contained in the food that they eat.
   c. In some ecosystems, the majority of the energy stored in plants does not become available until the plants die.
   d. Heat and energy are lost by each organism in the ecosystem.
   e. The two food webs are classified as grazing and detrital.

E   52. Energy flow in an ecosystem is
   a. cyclical.
* b. one-way.
   c. two-way.
   d. reversible under different conditions.

D   53. Assume that an energy pyramid has four levels (producers plus three consumers), and also assume that the energy in the producer level is set at 100%. What percent of the energy in the producers will be obtained by the tertiary consumers?
   a. 100
   b. 10
   c. 1
* d. 0.1
   e. 0.01

BIOGEOCHEMICAL CYCLES—AN OVERVIEW

E   54. Materials in sedimentary cycles
   a. pass through both a solid and a gaseous phase.
* b. are never present as gases in the ecosystem.
   c. are present as liquids in the earth but as gases in the atmosphere.
   d. pass through both a solid and a gaseous phase and are present as liquids in the earth but as gases in the atmosphere.

E   55. Which of the following does NOT cycle through an ecosystem?
   a. water
   b. carbon
* c. energy
   d. phosphorus
   e. nitrogen

M   56. The chemical elements that are available to producers are usually in what form?
* a. ions
   b. gases
   c. solids
   d. compounds
   e. hydrocarbons

D 57. Which of the following statements is false?
   a. Ecologists use models to represent relationships between biogeochemical cycles and most ecosystems.
 * b. The physical environment has virtually no reservoir for most elements.
   c. Inputs from the physical environment and recycling made possible by decomposers and detritivores maintain the nutrient reserves in an ecosystem.
   d. In most major ecosystems, the amount of nutrients that is cycled within the ecosystem is greater than the amount entering or leaving the ecosystem in a given year.
   e. Once elements are in the biological compartments of the biogeochemical cycles, they are unlikely to leave until the organism dies.

THE HYDROLOGIC CYCLE

M 58. The Hubbard Brook watershed studies revealed the importance of tree roots in preventing loss of calcium from an ecosystem. Calculation of calcium loss is performed by sampling
   a. the roots of the trees.
   b. the soil of the watershed.
 * c. the stream exiting the watershed.
   d. the roots of the trees and the soil of the watershed.
   e. the roots of the trees, the soil of the watershed, and the stream exiting the watershed.

M 59. Most of the water vapor in the earth's atmosphere comes from evaporation from
   a. lakes.
   b. rivers.
   c. land.
 * d. oceans.
   e. plants.

THE CARBON CYCLE

M 60. Which gas is increasing in the atmosphere and threatening the world with the greenhouse effect?
 * a. carbon dioxide
   b. carbon monoxide
   c. ozone
   d. fluorocarbons
   e. oxygen

E 61. Carbon is stored in what form?
   a. biomass
   b. fossil fuels
   c. limestone rocks
   d. shells of animals
 * e. all of these

M 62. Carbon is introduced into the atmosphere by all of the following means EXCEPT
   a. respiration.
   b. volcanic eruptions.
   c. burning of fossil fuels.
 * d. wind erosion.

D 63. Carbon enters the biomass of animal bodies in the form of
   a. carbon dioxide.
   b. carbon monoxide.
 * c. carbohydrates.
   d. fossil fuels.
   e. calcium carbonate.

THE NITROGEN CYCLE

M 64. Which is NOT part of the nitrogen cycle?
   a. denitrification
 * b. deammonification
   c. nitrogen fixation
   d. ammonification
   e. assimilation and biosynthesis

M 65. A process in which nitrogenous waste products or organic remains of organisms are decomposed by soil bacteria and fungi that use the amino acids being released for their own growth and release the excess as NH_3 or NH_4 is
   a. nitrification.
 * b. ammonification.
   c. denitrification.
   d. nitrogen fixation.
   e. hydrogenation.

E 66. The greatest concentration of nitrogen on the planet Earth is found in
   a. living organisms, including bacteria.
 * b. the atmosphere.
   c. soil minerals.
   d. fossil fuels.
   e. oceans.

E　67. Nitrogen is released into the atmosphere by
　　　a. nitrogen fixation.
　*　b. denitrification.
　　　c. nitrification.
　　　d. ammonification.
　　　e. decomposition.

E　68. Which plants are planted to increase the amount of nitrogen in the soil?
　　　a. watermelon and cantaloupe vines
　*　b. legumes
　　　c. mints
　　　d. grasses
　　　e. heaths

M　69. Plant cells assimilate nitrogen in the form of
　　　a. ammonia and N_2.
　　　b. N_2 and nitrite.
　*　c. nitrate and ammonia.
　　　d. urea and nitrate.

M　70. Nitrifying bacteria convert
　*　a. ammonia to nitrite and nitrate
　　　b. nitrate to nitrite to nitrogen gas
　　　c. nitrate to ammonia
　　　d. urea to ammonia.

D　71. Humans incorporate nitrogen into their bodies
　　　a. by nitrification.
　　　b. through ammonification.
　*　c. by assimilation and biosynthesis.
　　　d. via denitrification.
　　　e. by the process of decomposition.

THE PHOSPHORUS CYCLE

D　72. Animals obtain minerals such as phosphorus
　　　a. primarily dissolved in drinking water.
　　　b. by inhalation.
　　　c. in meats.
　　　d. by eating plants.
　*　e. by eating plants and meats.

FOCUS ON OUR ENVIRONMENT: TRANSFER OF HARMFUL COMPOUNDS THROUGH ECOSYSTEMS

E　73. In biological magnification
　*　a. poisons build up in food chains and webs so that the concentration is highest at the high end of the food chain.
　　　b. there is a tendency for an environment to change when organisms first invade.
　　　c. more highly evolved forms are able to build large populations under favorable conditions.
　　　d. parasites spread rapidly through congested populations.
　　　e. sediments fill in aquatic environments so that succession will occur if organisms disturb the aquatic habitat.

M　74. Which substance is magnified during transfers in ecosystems?
　*　a. fat soluble pesticides
　　　b. carbohydrates
　　　c. inorganic phosphates
　　　d. fat soluble pesticides and carbohydrates
　　　e. fat soluble pesticides, carbohydrates, and inorganic phosphates

M　75. Biological magnification refers to the
　　　a. increase in size of animals as they progress through a food chain.
　　　b. increase in size of organisms as they progress through ecological succession.
　　　c. increase in the efficiency of energy utilization as organisms progress through a food chain.
　*　d. accumulation of toxic pollutants as animals pass through a food chain.

M　76. The release of DDT into the environment to control some insect pests will result in the highest detectable concentrations
　　　a. at the bottom of the food chain.
　　　b. in the targeted insect pest.
　　　c. in the middle of the food chain.
　*　d. at the end of the food chain.

M　77. Long-lasting pesticides such as DDT
　　　a. are target-specific.
　　　b. are the most effective control over pests.
　*　c. build up in concentration as they pass through a food chain.
　　　d. break down after the organism that receives the pesticide dies.

Matching Questions

D   **78.** Choose the one most appropriate answer for each.

1 _____ biological magnification

2 _____ detritus

3 _____ legumes

4 _____ net primary production

5 _____ primary productivity

6 _____ webs

7 _____ gross primary production

A. rate at which energy becomes stored in organic compounds through photosynthesis

B. total amount of solar energy stored in organic compounds during photosynthesis

C. interconnected food chains

D. the potential chemical energy remaining (after aerobic respiration by autotrophs) that can still be passed on to other trophic levels

E. DDT spraying program in Borneo

F. a kind of plant that often harbors symbiotic nitrogen fixers in its roots

G. particles of organic waste products, dead or partly decomposed tissues

Answers:

1. E	2. G	3. F
4. D	5. A	6. C
7. B		

Classification Questions

Answer questions 79–83 in reference to the five trophic categories of an ecosystem listed below:

   a. producer
   b. herbivore
   c. primary carnivore
   d. secondary carnivore
   e. decomposer

M   **79.** This is a primary consume.

M   **80.** A Venus flytrap plant obtains its nitrogen when it functions as this.

M   **81.** Most mushrooms function in this capacity.

D   **82.** A bear feeding on a salmon is functioning as this.

E   **83.** A bear feeding on blueberries is functioning as this.

Answers: 79. b   80. c   81. e
                82. d   83. b

Answer questions 84-88 in reference to the four steps of the nitrogen cycles listed below:

   a. nitrogen fixation
   b. nitrification
   c. denitrification
   d. ammonification

M   **84.** The action of bacteria on urea occurs during this process.

M   **85.** The action of bacteria on ammonia, ultimately converting it to nitrate, occurs during this process.

E   **86.** The action of bacteria on nitrates, converting them to gaseous nitrogen, occurs during this process.

E   **87.** This is the process whereby gaseous nitrogen is first converted to ammonia and then to other nitrogenous compounds.

M   **88.** The process whereby nitrite is converted to nitrate is an important part of this process.

Answers: 84. d   85. b   86. c
                87. a   88. b

Selecting the Exception

E   89. Three of the four answers listed below are related by a common theme. Select the exception.
    a. numbers
    *   b. nutrients
    c. biomass
    d. energy

M   90. Four of the five answers listed below are heterotrophic. Select the exception.
    a. consumers
    b. carnivores
    c. herbivores
    d. parasites
    *   e. producers

M   91. Four of the five answers listed below are related by a common action. Select the exception.
    a. volcanic eruption
    *   b. photosynthesis
    c. respiration
    d. fire
    e. decomposition

D   92. Four of the five answers listed below are related by a common action that retains nitrogen in the biomass. Select the exception.
    a. decomposition
    b. ammonification
    c. nitrification
    d. nitrogen fixation
    *   e. denitrification

CHAPTER 25
IMPACTS OF THE HUMAN POPULATION

Multiple-Choice Questions

CHARACTERISTICS OF POPULATIONS

M **1.** The average number of individuals of the same species per unit of surface area at a given time is the
  * a. population density.
  b. population growth.
  c. population birth rate.
  d. population size.
  e. carrying capacity.

M **2.** To a person studying utilization of classroom space on campus, the most useful data concerning students in a classroom would be expressed by the number of
  a. total individuals.
  * b. individuals per square yard.
  c. individuals per room.
  d. rooms per building.
  e. students of each age.

M **3.** The distribution of the human population in the United States is
  * a. clumped.
  b. random.
  c. uniform.
  d. constant.

E **4.** Population size depends upon
  a. deaths.
  b. births.
  c. migration.
  d. immigration.
  * e. all of these

M **5.** The age structure diagram for rapidly growing populations
  a. is in the form of a pyramid.
  b. is characterized by a large percentage of the population in the postreproductive years.
  c. has a very broad base showing a large number of young.
  d. has about equal distribution between all age groups.
  * e. is in the form of a pyramid and has a very broad base showing a large number of young.

D **6.** If reproduction occurs early in the life cycle, what occurs?
  * a. population growth rate increases
  b. population size declines
  c. population size is not affected
  d. generation time increases
  e. growth rate remains unchanged

M **7.** As the twenty-first century draws near, the continent with the highest potential for human population growth rate was __?__ and that with the lowest was __?__
  a. North America, Asia
  b. Asia, North America
  * c. Africa, Europe
  d. Africa, Asia

M **8.** A situation in which the birth rate equals the death rate is called
  a. an intrinsic limiting factor.
  b. exponential growth.
  c. saturation.
  * d. zero population growth.
  e. geometric growth.

D **9.** The rate of increase for a population (r) refers to what kind of relationship between birth rate and death rate?
  a. their sum
  b. their product
  c. the doubling time between them
  * d. the difference between them
  e. reduction in each of them

E   10. Which characteristic of a population is a convenient way to express the rate of change within a population?
         a. size
     *   b. growth
         c. density
         d. carrying capacity
         e. age

E   11. A population that is growing exponentially in the absence of limiting factors can be illustrated by which curve?
         a. S-shaped
     *   b. J-shaped
         c. one that terminates in a plateau phase
         d. bimodal
         e. binomial

M   12. Which concept is a way to express the growth rate of a given population?
     *   a. doubling time
         b. population density
         c. population size
         d. carrying capacity
         e. all of these

M   13. The most reasonable method of limiting human population growth is
         a. increasing carrying capacity.
     *   b. decreasing birth rate.
         c. decreasing competition.
         d. increasing death rate.
         e. exploiting outer space.

D   14. In a population growing exponentially,
         a. the number of individuals added to the population next year is greater than the number added this year.
         b. the population growth rate increases year after year.
         c. net reproduction per individual increases year after year.
     *   d. the number of individuals added to the population next year is greater than the number added this year and the population growth rate increases year after year.

A CLOSER LOOK AT GROWTH PATTERNS

M   15. The maximum rate of increase of a population is its
     *   a. biotic potential.
         b. carrying capacity.
         c. exponential growth.
         d. distribution.
         e. reproductive base.

D   16. The biotic potential
         a. varies from one species to another.
         b. is controlled by the timing of the first reproduction.
         c. is controlled by the frequency of reproduction.
         d. is controlled by the number of offspring produced.
     *   e. all of these

M   17. Interaction between resource availability and a population's tolerance to prevailing environmental conditions defines
     *   a. the carrying capacity of the environment.
         b. exponential growth.
         c. the doubling time of a population.
         d. density-independent factors.

M   18. Populations
         a. are limited by only one factor at a time.
         b. increase arithmetically.
         c. increase indefinitely.
     *   d. are limited by the carrying capacity.
         e. are represented by a minimum of two different sizes.

D   19. A J-shaped growth curve is converted to an S-shaped one
         a. when the parents are past reproductive age.
         b. if the data are plotted in reverse.
     *   c. when the carrying capacity is reached.
         d. if reproduction stops.
         e. only for fast-growing populations such as bacteria.

D   20. The carrying capacity of an environment is determined by
         a. the net rate of reproduction of the female members.
         b. an S-shaped curve.
         c. the predation rate on the females.
         d. diseases suffered by both sexes.
     *   e. the sustainable supply of resources it provides.

D 21. In natural communities some feedback mechanisms operate whenever populations change in size; they are
  * a. density-dependent factors.
  b. density-independent factors.
  c. always within the individuals of the community.
  d. always outside the individuals of the community.
  e. none of these

M 22. A change in a population that is NOT related strictly to the size of the population is best described as
  a. density-dependent.
  * b. density-independent.
  c. within.
  d. an S-shaped curve.
  e. a J-shaped curve.

M 23. In itself, a flood that washes away an entire population of rabbits is
  a. a density-dependent factor.
  b. a limiting factor dependent on the individuals.
  c. a consequence of exponential growth.
  * d. density-independent.
  e. all of these

M 24. Which density-dependent factor controls the size of a population?
  a. wind velocity
  b. light intensity
  * c. nutrient supply
  d. rainfall
  e. wave action in an intertidal zone

E 25. Which is NOT a density dependent, growth-limiting factor?
  a. predation
  * b. drought
  c. parasitism
  d. competition

E 26. Life tables provide data concerning
  a. expected life span.
  b. reproductive age.
  c. death rate.
  d. birth rate.
  * e. all of these

HUMAN POPULATION GROWTH

E 27. Which of the following is closest to the actual birth rate for humans in 1999?
  a. 1 billion per year
  * b. 273,000 per day.
  c. 1.9 million per day.
  d. 100 million per month.
  e. 100,000 per hour.

M 28. In which stage of the demographic transition model is population growing the fastest?
  a. preindustrial
  * b. transitional
  c. industrial
  d. postindustrial

M 29. In which stage of the demographic transition model is zero population a reality?
  a. preindustrial
  b. transitional
  c. industrial
  * d. postindustrial

E 30. In which stage of the demographic transition model is the country of Mexico at the close of the twentieth century?
  a. preindustrial
  * b. transitional
  c. industrial
  d. postindustrial

E 31. In which stage of the demographic transition model is the United States at the close of the twentieth century?
  a. preindustrial
  b. transitional
  * c. industrial
  d. postindustrial

AIR POLLUTION

M 32. In the United States, how many metric tons of pollutants are discharged into the atmosphere each day?
  a. 1,000
  b. 100,000
  * c. 700,000
  d. 5 million
  e. 15 to 30 million

E 33. Air pollution
    a. reduces visibility.
    b. corrodes buildings.
    c. causes various human diseases.
    d. damages plants.
*   e. all of these

E 34. Air pollution may cause
    a. lung cancer.
    b. emphysema.
    c. bronchitis.
    d. burning eyes.
*   e. all of these

E 35. Transportation-produced smog causes air to turn
    a. gray.
    b. black.
*   c. brown.
    d. red.
    e. blue.

E 36. Insustrial smog would be the result of which of the following?
    a. burning fossil fuels
    b. particulates in the air
    c. a thermal inversion.
*   d. all of these

E 37. Industrial smog causes air to turn
*   a. gray.
    b. black.
    c. brown.
    d. red.
    e. blue.

M 38. Nitric oxide in the exhaust of cars and trucks results in
*   a. photochemical smog
    b. industrial smog
    c. a thermal inversion
    d. all of these

M 39. Which factor is NOT characteristic of a city primarily plagued by industrial smog?
    a. high concentration of sulfur oxides
    b. fossil fuels used for manufacturing
    c. cold, wet winters
*   d. high concentration of nitrogen oxides

M 40. Which factor is NOT characteristic of a city primarily plagued by photochemical smog?
*   a. high concentration of sulfur oxides
    b. high concentration of nitrogen oxides
    c. significant amounts of PANs
    d. large numbers of internal combustion engines

M 41. In brown air fog, which substance combines with nitrogen dioxide in the sunlight to form photochemical smog?
    a. carbon monoxide
    b. water vapor
*   c. hydrocarbons
    d. sulfuric acid
    e. all of these

D 42. What results when nitrogen dioxide and hydrocarbons react in the presence of sunlight?
*   a. photochemical smog
    b. industrial smog
    c. a thermal inversion
    d. both photochemical smog and a thermal inversion

M 43. Which acid is a severe air pollutant?
    a. carbonic acid
*   b. nitric acid
    c. hydrofluoric acid
    d. hydrochloric acid
    e. boric acid

E 44. Acid rain
    a. attacks nylons.
    b. attacks marble statues.
    c. causes toxic metals to become motile in the ecosystem.
    d. can be reduced in the local area by tall smokestacks.
*   e. all of these

M 45. Each substance contributes to wet acid deposition EXCEPT
*   a. ozone.
    b. waste products from the burning of coal.
    c. nitrogen fertilizers.
    d. waste products from the burning of gasoline.

M 46. The unequal distribution of acid rain over the United States is closely correlated with
    a. per capita energy use.
    b. fertilizer use.
*   c. burning fossil fuels.
    d. average summer temperatures.

M 47. The region of the United States most affected by acid rain is the
    a. Northwest.
    b. Southwest.
*   c. Northeast.
    d. Southeast.

M 48. The atmosphere above which region is known to have a hole in the ozone layer?
* a. Antarctica
b. Eastern North America
c. Northern Europe
d. the western Pacific

M 49. Each factor appears to be correlated with a decrease in atmospheric ozone EXCEPT
a. suppression of the immune system.
b. decreased rates of photosynthesis.
c. increased incidence of skin cancers.
* d. decreased levels of atmospheric carbon dioxide.

M 50. Uses of chlorofluorocarbons include each of the following EXCEPT
* a. gasoline additives.
b. aerosol propellants.
c. refrigeration coolants.
d. plastic packaging.

M 51. Chlorofluorocarbons (CFCs) are pollutants because
* a. biogeochemical mechanisms for their removal have not yet appeared in the biosphere.
b. they combine with water to form hydrochloric and hydrofluoric acids.
c. they are found in smog.
d. they are photochemical oxidants.

A GLOBAL WATER CRISES

E 52. For every million liters of water in the world, only about _____ liters are in a form that can be used for human consumption or agriculture.
* a. 6
b. 60
c. 600
d. 6,000

M 53. Irrigated land accounts for approximately what percentage of human food production?
a. 10
* b. 30
c. 50
d. 75

M 54. Which is NOT a result or effect of irrigation?
a. increased food production
b. waterlogging of soil
c. lowered water tables
* d. uplifted land
e. salinization

WHERE TO PUT SOLID WASTES? WHERE TO GROW FOOD?

E 55. About how many beverage containers sold in the United States each year are nonreturnable cans and bottles, many of which are discarded in public places?
a. 25 million
b. 25 billion
c. 50 million
* d. 50 billion
e. 100 million

E 56. What percentage of urban wastes are paper products?
a. 10
b. 20
* c. 50
d. 75
e. more than 90

E 57. Using recycled paper could reduce the air pollution that results form paper manufacturing by
a. 50 percent
* b. 95 percent
c. 5 percent
d. zero percent.
e. 75 percent

E 58. Landfills should contain
a. all garbage.
b. nonbiodegradable wastes.
* c. nonrecycled solid wastes.
d. glass and metallic debris.
e. organic litter.

E 59. How many trees are required just to print all the Sunday newspapers in the USA?
a. 1 million
b. 100,000
* c. 500,000
d. 2 billion
e. none because all the paper is recycled

M 60. Approximately 50 percent of the billions of tons of solid wastes produced in the United States is
a. glass.
* b. paper.
c. aluminum.
d. plastic.

M 61. Of the earth's land, what is the maximum percentage now being used for agriculture?
  a. 10
  * b. 21
  c. 50
  d. 75

M 62. The new high-yield crops require which of the following that cannot be supplied by subsistence agriculture?
  a. irrigation
  b. pesticides
  c. fertilizers
  d. fossil fuel energy
  * e. all of these

M 63. The primary cause of desertification in the world today is
  a. increased salinity resulting from irrigation practices.
  * b. overgrazing of marginal lands.
  c. clearing and degradation of tropical forests.
  d. herbicide and fertilizer runoff.

DEFORESTATION—AS ASSAULT ON FINITE RESOURCES

M 64. Deforestation results in
  a. increased air temperatures.
  b. decreased soil fertility.
  c. decreased transpiration rates of individual plants.
  * d. increased air temperatures and decreased soil fertility.
  e. increased air temperatures, decreased soil fertility, and decreased transpiration rates of individual plants.

M 65. At present rates of clearing and degradation, the disappearance of the tropical rain forest biome will be complete by the year
  * a. 2035.
  b. 2100.
  c. 2150.
  d. 2200.

CONCERNS ABOUT ENERGY USE

E 66. Fossil fuels
  a. are renewable natural resources.
  * b. will be commercially depleted within the next 100 years.
  c. can be utilized without any environmental degradation.
  d. are essentially pure carbon deposits.

M 67. If one kind of plutonium isotope is not removed, the wastes from a nuclear power reactor must be kept out of the environment for how many years before they are safe?
  a. 25
  b. 250
  * c. 250,000
  d. 1 million
  e. 250 million

M 68. Which statement about nuclear power plants is true?
  * a. Their net energy production is relatively low.
  b. Their waste products lead to the production of acid rain.
  c. Their waste products are not radioactive.
  d. Their net energy production is relatively low; and Their waste products lead to the production of acid rain.

M 69. Problems associated with the extraction or use of coal as a large-scale source of energy include all of the following EXCEPT
  a. production of sulfur oxides.
  b. stripmining in fragile semiarid environments.
  * c. limited reserves.
  d. amplification of the general warming trend of the earth.

E 70. Which of the following shows the most promise as an energy source in the next century?
  a. breeder reactors
  * b. solar-hydrogen energy
  c. wind energy
  d. fusion power
  e. fossil fuels

THE IDEA OF SUSTAINABLE LIVING

E   **71.** All of the following are incompatible with the idea of a sustainable society except
    a. reliance on fossil fuels for energy needs.
    b. continued deforestation.
    c. overfishing of the oceans.
    * d. reducing the birth rate.
    e. continued production of CFCs.

Matching Questions

D   **72.** Choose the one most appropriate answer for each.

1 _____ age structure

2 _____ population growth rate

3 _____ J-shaped curve

4 _____ limiting factor

    A. describes a population that is experiencing unrestrained growth

    B. the birth rate minus the death rate plus any inward or minus any outward migration

    C. how individuals are distributed at each age level for a population

    D. the amount of glucose in a culture flask containing bacteria

Answers: 1. C   2. B   3. A
                 4. D

Classification Questions

Answer questions 73-77 in reference to the five terms listed below:

    a. carrying capacity
    b. net reproductive rate
    c. age structure
    d. survivorship
    e. growth rate

D   **73.** This is the average number of offspring born to each female over her reproductive lifetime.

M   **74.** This is the relative proportion of individuals of each is age.

M   **75.** This is the maximum number of individuals that a given habitat can support.

D   **76.** A change in the available food supply in a habitat will affect all of these.

E   **77.** This term is equal to the birth rate minus the death rate of a species.

Answers: 73. b   74. c   75. a
            76. c   77. e

Answer questions 78-81 in reference to the four terms listed below:

    a. preindustrial stage
    b. transitional stage
    c. industrial stage
    d. postindustrial stage

D   **78.** In this stage zero population growth is reached.

M   **79.** In this stage there is little population growth because the conditions are so harsh that death rates are high.

M   **80.** In this stage does industrialization begins, death rates drop, but birth rates remain high.

D   **81.** During this stage population growth slows due to stable industrialization.

Answers: 78. d   79. a   80. b
            81. c

Selecting the Exception

E **82.** Four of the five answers listed below are factors that affect population size. Select the exception.
   a. births
 * b. distribution
   c. emigration
   d. immigration
   e. deaths

M **83.** Four of the five answers listed below are density-independent factors. Select the exception.
 * a. nutrient supply
   b. temperature drop
   c. drought
   d. volcanic eruption
   e. hardfreeze

D **84.** Four of the five answers below are limits to achieving a population's full biotic potential. Select the exception.
   a. lack of nutrients
   b. predation
 * c. exponential growth
   d. competition for space
   e. waste buildup

E **85.** Four of the five answers listed below are particulate wastes. Select the exception.
   a. smoke
   b. soot
   c. asbestos
   d. dust
 * e. ozone

E **86.** Four of the five answers listed below are nonrenewable resources. Select the exception.
 * a. biomass
   b. coal
   c. natural gas
   d. oil
   e. nuclear

M **87.** Four of the five answers listed below are effects of acid rain. Select the exception.
   a. attacks marble, metals, mortar, nylons
 * b. causes air inversion and pollution events
   c. makes toxic heavy metals more mobile
   d. has different effects in different watersheds
   e. produces sterile lakes

M **88.** Four of the five answers listed below are human efforts to improve the carrying capacity of the earth for humans. Select the exception.
 * a. desertification
   b. recycling
   c. green revolution
   d. irrigation
   e. conservation